A NATURALIST'S
BOOK OF
WILDFLOWERS

A NATURALIST'S BOOK OF WILDFLOWERS

Celebrating 85 Native
Plants of North America

LAURA C. MARTIN

THE COUNTRYMAN PRESS
A DIVISION OF W. W. NORTON & COMPANY
INDEPENDENT PUBLISHERS SINCE 1923

This volume is intended as a general information resource. Neither the publisher nor the author can guarantee that every reader will be able accurately to identify every plant described in these pages. Please use common sense before gathering or coming into other contact with plants that may cause an allergic reaction and do not ingest anything that you have not definitively identified as safe and non-toxic. Use particular care when cooking or crafting with children. This book is not a substitute for medical advice; herbal remedies should not be used in lieu of individualized attention from a qualified professional. As of press time, the URLs displayed in this book link or refer to existing websites. The publisher is not responsible for, and should not be deemed to endorse or recommend, any website other than its own or any content available on the Internet (including, without limitation, any website, blog page, or information page) that is not created by W. W. Norton. The author, similarly, is not responsible for third-party material.

FOR MY GRANDCHILDREN:
RIVERS, ELLIE, ROWAN,
BRAXTON, DAVEY, AND GEORGE

CONTENTS

INTRODUCTION

I have spent my entire life loving wildflowers. I grew up in the woods on the outskirts of the city, and because my mother was a wildflower enthusiast, I became one too. I eventually acquired a bachelor's degree in botany and then a certification in botanical illustration. The result is *A Naturalist's Book of Wildflowers: Celebrating 85 Native Plants of North America*. This "celebration" comes through scientific text, interesting folklore, detailed botanical drawings, and whimsical sketches. As both a trained botanist and a botanical illustrator, I have learned to look at plants from many different angles, and I am happy to share this perspective with you. I hope that this book provides you with a fun, fascinating, and enjoyable way to examine and celebrate our native plant treasures.

In the plant entries, I have included annotated drawings, time of bloom, natural range, conservation status, wildlife partners, medicinal uses, gardening tips, and lots of fun folklore, superstitions, and trivia.

The wildflowers are, in my humble opinion, the most beautiful part of an ecosystem. They are no more important than any other element, but I think they are the most beautiful, making their study even more pleasurable. But in choosing which plants to include in this book, I wanted to look beyond physical beauty and consider, also, the fascinating cultural and natural histories of each plant. For example, Indian hemp is easily overlooked in the field. With small, plain white flowers, it simply can't compare to the beauty of a glacier lily or a columbine. But a little research uncovered the fact that Indian hemp was an incredibly useful plant for many American Indian tribes, as they used the inner pith for making string and cord.

Many native plants provided essential resources for American Indians. Plants were used for food, medicine, and ceremony. They were used for weaving, cordage, dyes and inks, and much more. Sometimes the existence of a tribe depended on the wildflowers growing around them. For example, bitterroot was such an important food source for the Kootenai and Flathead that it was used as currency. You could buy a horse for a 50-pound bag of bitterroot.

The indigenous peoples of North America found medicines in the woods and fields, plains and deserts in which they lived. Yarrow was used to staunch blood flowing from a wound, trillium was used during childbirth, and bloodroot was used (sparingly, due to its toxic nature!) to treat colds and congestion. When settlers from the eastern United States began to move west, they too became dependent on the native plants.

I included other plants because of their essential role in nature. For example, milkweed is an irreplaceable food source for the larvae of monarch butterflies. Without milkweed, we'd have no monarchs. Native plants provide not only food and nectar for wildlife but also nesting sites and protection for myriad species, making them a necessary part of their natural community. It is not a one-sided relationship. Flowers are dependent on pollinators and, in some cases, such as the lady's slipper, on the fungus in the soil, which provides critical nutrients to the plant. In the individual plant entries, I have included information about these relationships under the heading "Wildlife Partners."

Some plants were included simply because they are amazing. Take, for example, the small bunchberry, which earned the nickname "firecracker" because it releases pollen at the startling rate of 24,000 meters per second. Our native plants truly are full of magic.

The United States is composed of many complex and overlapping ecosystems, and I have included plants that are indigenous to a wide variety of naturally occurring plant communities.

Most people, especially gardeners, are accustomed to thinking about plant distribution as it appears in USDA horticultural growing zones, which are based on average minimum temperatures of a region. These make up horizontal bands dividing the country from north to south. Ecologically, North America has been divided by the Commission for Environmental Cooperation into broad plant community zones that are based on dominant vegetation as well as soil type, moisture, wildlife, geology, and other plant forms. These are grouped more vertically than horizontally.

For example, the Eastern Temperate Forests make up a large and diverse ecosystem, and it should be no surprise that many of our native wildflowers are

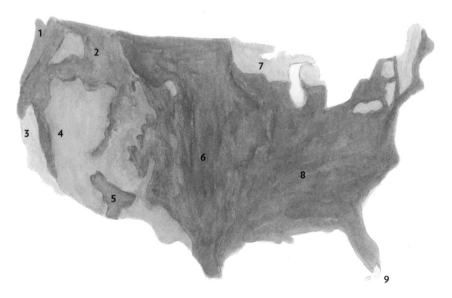

1. Marine West Coast Forest   2. Northwest Forested Mountains   3. Mediterranean California
4. Temperate Sierras   5. North American Deserts   6. Great Plains   7. Northern Forests
8. Eastern Temperate Forests   9. Tropical Wet Forests

found in this system. The West is a little more complex in that it features plant communities that are found in areas as diverse as deserts to high alpine meadows. The natural ranges and regions listed in this book are broad indications only and provide just a general idea of where these plants are found.

As an artist, it was impossible for me not to substitute drawings for words, particularly while out in the field. It was much easier for me to remember the plants with arrows and sketches and margin notes than to simply write about them. So I've included these drawings with the hope that through them, the personalities of these plants will come alive for you.

As lovely as it is to look and wonder (and maybe draw!) these plants, it's also tons of fun to use the more abundant ones for a variety of purposes. I've included chapters on gardening, foraging and cooking, wildflower crafts, and doing activities with kids. I've also included a little botany to help with identification and a lot of whimsy, folklore, superstition, and stories just because it's fun.

Throughout the ages, people have been fascinated with flowers: their uses, symbolism, and stories. In this book, I often refer to two ways people engaged with wildflowers. The first is the "language of flowers," created during the Victorian Age, where each flower came to symbolize an emotion, feeling, or value, from love (rose) to wisdom (columbine). The other is the "doctrine of signatures," dating back to the 17th century in which each flower was considered to be useful in curing whatever part of the body the flower physically resembled. For example, plants with heart-shaped leaves were often used to treat heart ailments.

It's been forty years since I published my first book, *Wildflower Folklore*. A lot has changed in that time. Some of the plants have been renamed or shuffled around to be placed in a different genus. Archeologists have continued to broaden our understanding of the importance of plants to our ancient ancestors. Scientists are looking toward the future to discover how our native wildflowers might help mitigate the effects of climate change.

One thing that has become very apparent to me over the last four decades is our understanding of how vital our native plants are to their ecosystems and how invasive, non-native species are threatening their very existence. For this reason, I have included only our beloved native species in this book.

Whether you simply observe our native plants through a car window as you speed down the highway, or you get up close and personal on your hands and knees in the woods, I hope that this book will provide the inspiration to fall in love with our natives and to do what you can to protect and preserve them. Our flowers and their communities are gifts of incomparable value, and it's our joyful responsibility to pass these on to future generations with love and respect.

(JUST A LITTLE) BOTANY

Although it's not essential to know botany to appreciate wildflowers, it might make it a little more fun and interesting. Botany is just the study of plants, so if you put a little effort into studying them, your appreciation will increase greatly.

I encourage you to learn the Latin names. In some instances, this will be easy—violets are *Viola,* gentians are *Gentiana,* iris are *Iris!* Other names (*Chamaenerion angustifolium,* fireweed) are not so simple. While common names are much easier to pronounce and remember, the same plant may have any number of common names, depending on where it grows, making it a little confusing to use just the common name.

LEAF SHAPES AND FLOWER PARTS

While there is an endless array of flower types and configurations, and multitudes of different kinds of leaves, they all tend to fall into broad categories that are fairly straightforward and easy to remember. Leaf shapes are named after, well, after their shapes: lance, oval, heart, and so forth. Flower parts are a little more complicated and can sometimes be very complicated (such as with orchids), but they still can be categorized and identified pretty easily. The placement of the sexual parts of the flowers, where and how the flowers appear on the plant, and the shape and arrangement of the leaves are all clues to identification.

The following can be used as a quick reference guide:

FLOWER PARTS

Sexual parts: anther, petal, style, stigma, stamens, sepals, ovary, receptacles

Composite flower: disk flowers, ray flowers

Corolla: all the petals (usually white or colored)

Calyx: all the sepals (usually green)

Bracts: modified leaves, often surrounding a flower, not part of a calyx

Lip and lobe: shape of petals and blossom

Tepals: petals and sepals that look identical

Pedicel: flower or fruit stalk

FLOWER ARRANGEMENTS

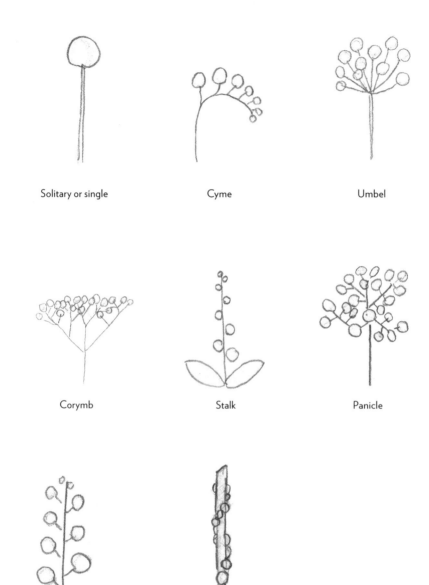

Solitary or single

Cyme

Umbel

Corymb

Stalk

Panicle

Raceme

Spiral

FLOWER SHAPES

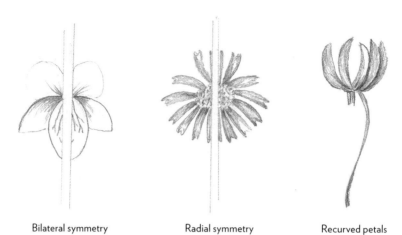

Bilateral symmetry   Radial symmetry   Recurved petals

Pea-like flower   Pea-like flower side view

LEAF ARRANGEMENTS

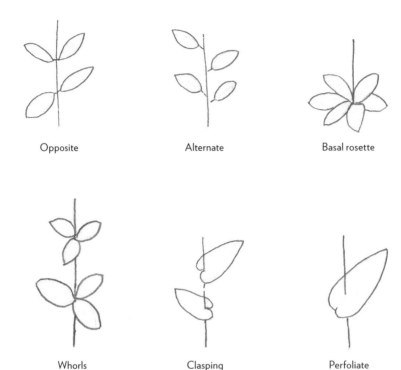

Opposite   Alternate   Basal rosette

Whorls   Clasping   Perfoliate

"Love is like wildflowers;
It's often found in the
most unlikely places."
—Ralph Waldo Emerson

ROOTS AND BULBS

Corm: underground stem

Bulb: rounded underground storage organ

Taproot: straight tapering root

Fibrous root: thin branching roots from stem

Tuberous roots: thickened underground roots

Rhizome: underground horizontal stem, usually sends out roots

YELLOW AND BROWN FLOWERS

Arnica, heartleaf *Arnica cordifolia*

Balsamroot, arrowleaf *Balsamorhiza sagittata*

Bellwort *Uvularia grandiflora*

Black-eyed Susan *Rudbeckia hirta*

Buttercup, western *Ranunculus occidentalis*

Cactus, Great Plains prickly pear *Opuntia polyacantha*

California poppy *Eschscholzia californica*

Cattail *Typha latifolia*

Coreopsis, plains *Coreopsis tinctoria*

Evening primrose *Oenothera biennis*

Goldenrod *Solidago* species

Jack-in-the-pulpit *Arisaema triphyllum*

Jewelweed, spotted touch-me-not *Impatiens capensis*

Lily, glacier *Erythronium grandiflorum*

Skunk cabbage, American *Lysichiton americanus*

Skunk cabbage, eastern *Symplocarpus foetidus*

Sunflower *Helianthus annuus*

ARNICA, HEARTLEAF

Arnica cordifolia

TOXIC

4-24 inches tall

flowers 2-3½ inches across

1-5 blossoms per plant

hairy bracts underneath flowers

10-15 yellow ray flowers

2-4 pairs of leaves with petioles

yellow-orange disk flowers

leaves 1½-5" long

heart-shaped roughly-toothed leaves

BLOOMS: May–September.

HABITAT AND RANGE: Open fields, meadows, wood edges, and lightly shaded woods. Can grow to 11,000 feet elevation. Found throughout the West, Western Canada, and around the Great Lakes region.

CONSERVATION: There are 28 species of Arnica native to North America, most of which are locally abundant. Common leopard's bane (*Arnica acaulis*), is endangered in Florida. Longleaf arnica (*A. lonchophylla*) is a threatened species in Minnesota. The European species, *A. montana* has been over-harvested in the wild and is now on the International Union for Conservation Nature (ICUN) Red List of Threatened Species.

WILDLIFE PARTNERS: This is an essential grazing food for the mule deer and comprises about a quarter of its diet during summer months. It is also eaten by elk, pronghorn sheep, upland game birds, and small mammals. Many different insects, including native bees and moths, pollinate it. It is an important host plant for the moth, *Bucculatrix arnicella*.

IN THE GARDEN: Growing arnica is possible almost anywhere, though it prefers higher elevations. It needs moist, well-drained soils, full sun or partial shade. It is a great plant to include in a pollinator garden, particularly in its native region. Plants live 12 years or more.

RELATED SPECIES: Leopard's bane or wolf's bane (*Arnica montana*).

MEDICINAL USES: Arnica montana has been cultivated and used as a healing herb in Europe since the 1500s. It has been harvested from the wild to be used as a major ingredient in salves. Although our native heart-leaf arnica does not offer as effective a medicine as its European cousin, the properties are similar. Both leaves and flowers are used by herbalists to externally treat swelling and sprains, and increase circulation. Because *all parts of the plant are considered toxic*, do not use this on open wounds. The native species was apparently not used medicinally by any American Indian peoples. *Toxic.*

The genus name is from the Greek word *arni,* meaning lamb, referring to the soft, downy leaves. According to legend, the medicinal properties of the European species were discovered by shepherds who noticed that injured sheep and goats were attracted to the plant. They believed that the animals sought out and ate the plants to soothe their injuries. The shepherds began using it themselves as a medicinal herb.

The species name comes from the Latin words for heart, *cord,* and leaf, *folia,* referring to the shape of the leaf. Another common name, mountain tobacco, was used because some thought that the leaf looked something like a tobacco leaf. It is, like its European cousin, also called leopard's bane.

In spite of its toxicity, small amounts of the plant have been occasionally safely used as flavoring in beverages, candy, and puddings, though this is not recommended. Arnica is also used in hair tonic and anti-dandruff products, and in cosmetics.

BALSAMROOT, ARROWLEAF

Balsamorhiza sagittata

8-32 inches tall

bright yellow ray flowers (8-25)

each blossom 2-4 inches across on leafless stalk

bracts quite hairy

American painted lady butterfly

leaves silvery gray, 12 inches or more long, from base of plant

large arrow-shaped leaves with petiole from basal cluster

disk flowers, each encased in papery scale

BLOOMS: May–July.

HABITAT AND RANGE: Grasslands, open woods, and hillsides from British Columbia east to Montana, south to Colorado, and west to California.

CONSERVATION: Abundant throughout the range.

WILDLIFE PARTNERS: This is an important forage plant for elk, bighorn sheep, and mule deer. It provides pollen for mason, mining, long-horned, sweat, and cuckoo bees, and many butterflies, including checkerspots, swallowtails, and American painted lady. Deer, mice, red-tailed chipmunk, ground squirrel, partridge, and sage grouse eat the seeds.

IN THE GARDEN: This plant needs a dry, sunny spot with room to spread. It survives drought and is long-lived, often surviving up to 40 years.

MEDICINAL USES: The Blackfoot used the leaves to make a poultice for wounds, cuts, bruises, and burns. The Cheyenne made a tea from the root to treat coughs and congestion.

This was an important survival plant for the Blackfoot and Cheyenne. In extreme need, they chewed on the stems to reduce both thirst and hunger. The Flathead, Nez Perce, and Kootenai peeled the young flower stems and ate the tender stalk. They also ate the root after baking it in a fire for three days or grinding it into a flour-like powder for making biscuits or small cakes. The seeds were eaten raw, roasted, or mixed with animal fat and made into small balls and used for "travel" food. The oil from the seeds was used for cooking.

Before the Plains Indians had horses, the Cheyenne and other tribes sent runners out to force buffalo to stampede over a cliff. They believed that smoke from burned balsamroot helped these runners who sometimes ran 20 miles or more.

The large, woody roots contain a lot of resin and grow 3 feet deep and are 4 to 5 inches in diameter. The deep roots can survive forest fires. After a fire, the number of balsamroots actually increases.

Balsamroot closely resembles several other plants growing at the same time and in the same general region. It can be easily identified with close observation. Heartleaf arnica (*Arnica cordifolia*) flower heads are smaller; the smooth leaves are on flowering stems. Mule's ear (*Wyethia amplexicaulis*) has shiny, waxy, large leaves (without stems or petioles) that grow from clumps at the base of the plant.

From left to right: Mule's ear, balsamroot, arnica

BELLWORT

Uvularia grandiflora

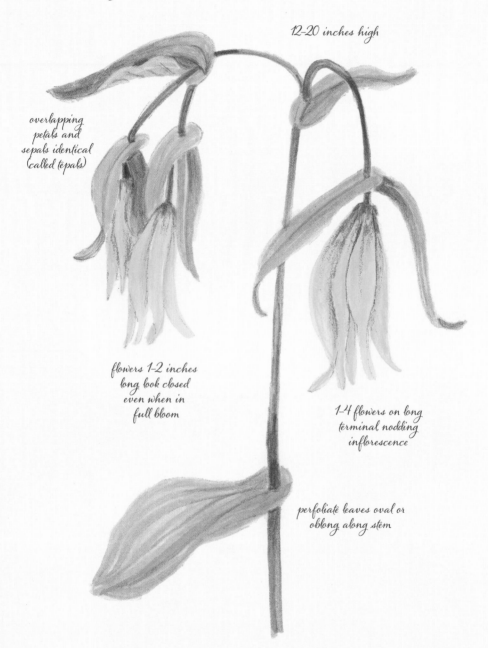

12–20 inches high

overlapping
petals and
sepals identical
(called tepals)

flowers 1–2 inches
long, look closed
even when in
full bloom

1–4 flowers on long
terminal nodding
inflorescence

perfoliate leaves oval or
oblong along stem

BLOOMS: April–June. After blooming, leaves persist and the plant continues to grow.

HABITAT AND RANGE: *U. grandiflora* grows in moist woods from Quebec and New England west to North Dakota, south to Louisiana, and east to Florida.

CONSERVATION: This plant is endangered in New Hampshire and Connecticut.

WILDLIFE PARTNERS: This is host to the rare bellwort mining bees (*Andrena uvulariae*), which live only off bellwort pollen. Bellwort is pollinated by bumblebees, halictid bees, and andrenid bees. Foliage is beloved by deer.

IN THE GARDEN: Bellwort needs shade, moist, fertile soil, and medium watering. Plant in clumps.

RELATED SPECIES: Other North American species include mountain bellwort (*Uvularia puberula*), perfoliate bellwort (*U. perfoliata*), Florida bellwort (*U. floridana*), and sessile bellwort (*U. sessilifolia*).

MEDICINAL USES: Many American Indian tribes made a tea for throat problems, canker sores, and mouth infections. The Menominee made a poultice to treat swelling, skin rashes, and wounds. The Potawatomi mixed the root with animal fat to use as a salve for sore muscles, backaches, and chest pains.

This, like many other plant names, ends in "wort." It comes from *wyrt,* meaning root or herb, and indicating that it had medicinal or culinary value. The genus comes from the word *uvula,* meaning "little grape."

Other common names include great merrybells or fairybells.

According to the doctrine of signatures, a theory proposed in the 17th century that suggested that a plant's physical appearance was a clue to its medicinal use, bellwort was used historically in Europe to treat throat ailments. Bellwort blossoms look like the uvula—the small pink thing that hangs down from the soft palate in the back of the throat.

Leaves and young shoots are tasty and are cooked like asparagus.

BLACK-EYED SUSAN

Rudbeckia hirta

12-36 inches tall

yellow ray flowers

bright yellow blossoms, 2-3 inches across

disk flowers reddish-brown (not true black)

lance-shaped leaves up to 8 inches long

sturdy, hairy stem

BLOOMS: June–September.

HABITAT AND RANGE: Commonly found along roadsides and in fields and open places. Native to the prairie states, this flower has now naturalized throughout the East and into the western states.

CONSERVATION: Common and abundant.

WILDLIFE PARTNERS: Black-eyed Susans are hosts for the gorgone checkerspot (*Chlosyne gorgone*) and bordered patch butterfly larvae. It is an important pollinator plant for all kinds of bees, wasps, beetles, flies, and many kinds of butterflies, including the buckeye. Several types of birds, such as goldfinch, chickadee, cardinal, nuthatch, sparrows, and towhees, eat the seeds, which persist through winter.

IN THE GARDEN: Hardy and easy to grow in full sun, the plant is adaptable and tolerates a variety of soil types. It has also been hybridized to form the gloriosa daisy. The cultivar 'Indian Summer' won a Royal Horticultural Society (RHS) Award of Garden Merit. *Rudbeckia hirta* 'Moreno" is particularly beautiful for the garden.

MEDICINAL USE: Modern testing found the plant contains antibodies. It was traditionally used by the Cherokee for skin infections, colds (tea made from leaves), flu, earaches, snake bites, sores, and swelling.

This genus was named for a Swedish botanist, Olof Rudbeck, who taught botany to the "father of modern botany," Carolus Linnaeus.

Black-eyed Susan is shunned by all types of livestock. There is some indication that it is somewhat toxic to cats.

The yellow flowers are used as a natural dye.

Black-eyed Susan is considered a symbol of justice. It is the state flower of Maryland.

BUTTERCUP, WESTERN

Ranunculus occidentalis

TOXIC

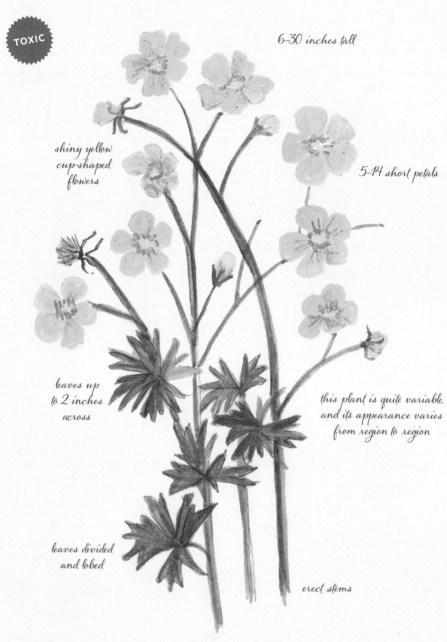

6–30 inches tall

shiny yellow
cup-shaped
flowers

5–14 short petals

leaves up
to 2 inches
across

this plant is quite variable,
and its appearance varies
from region to region

leaves divided
and lobed

erect stems

BLOOMS: March–August.

HABITAT AND RANGE: Found in moist, wetland areas, coastal prairies, meadows, and forests throughout western North America, from southern areas of California and north to Alaska.

CONSERVATION: Common and abundant. A naturalized buttercup from Northern Europe, the creeping buttercup (*Ranunculus repens*), is considered invasive in many areas.

WILDLIFE PARTNERS: The acrid taste of the leaves keeps this plant from being eaten by most wildlife. The pollen and nectar, though, are important for many different pollinators. It is primarily pollinated by native bees and pollinating flies. It does provide some butterfly larval food as well.

IN THE GARDEN: Easy to grow and good for attracting pollinators. Avoid planting the invasive Eurasian species.

The flower and sap of many species are poisonous and can severely irritate skin. One species of *Ranunculus* was used in making poison arrows.

Buttercups were important to the native peoples of California and Oregon because they believed the coming of the blooms marked the beginning of the salmon run.  They also used the seeds, ground and mixed with other seeds, to make pinole, a kind of flour (See page 243 for definition of pinole).

The Nez Perce of the Pacific Northwest called these flowers Coyote's Eyes, based on the legend that one day Coyote was playing with his eyeballs, tossing them into the air and catching them, when Eagle caught them and flew away with them. Coyote couldn't see but made new eyes from buttercups.

There are over 500 species of *Ranunculus* worldwide, many of which grow in swampy or boggy areas. This might account for the genus name, *Ranunculus,* which is Latin for "little frog."

Buttercups at some point acquired the reputation of causing lunacy and were called crazy weeds. Holding the flower up to your neck on the night of a full moon was said to drive you insane. An old country custom is to hold a buttercup under your chin. If your chin shines yellow, you love butter!

CACTUS, GREAT PLAINS PRICKLY PEAR

Opuntia polyacantha

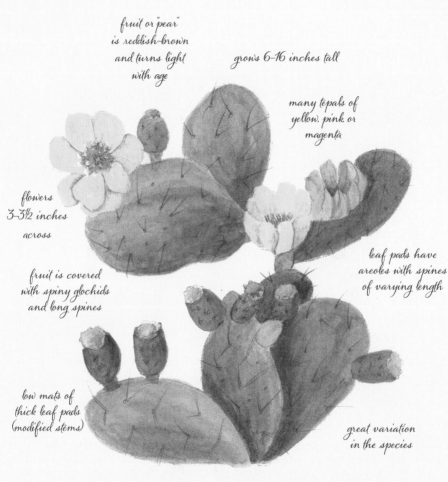

fruit or "pear" is reddish-brown and turns light with age

grows 6-16 inches tall

many tepals of yellow, pink or magenta

flowers 3-3½ inches across

leaf pads have areoles with spines of varying length

fruit is covered with spiny glochids and long spines

low mats of thick leaf pads (modified stems)

great variation in the species

mats can be 6-9 feet across

BLOOMS: April–May.

HABITAT AND RANGE: The most widespread of any North American cactus, it is found in every state west of the Mississippi River and in Canada and Mexico. It tolerates temperatures as low as -50°F and as high as 100°F. This plant grows in poor soils and can be found in a variety of habitats including chaparral, prairie, and wastelands.

CONSERVATION: Common in most of its range. Abundance indicates poor soils, as it outcompetes other vegetation.

WILDLIFE PARTNERS: Though the barbed spines protect it from most wildlife, the fruit provides about half the winter food source for black-tailed prairie dogs. If forest fires burn off the spines, it is also eaten by pronghorns.

IN THE GARDEN: Easy to grow in a desert garden. Beware of spines when cultivating, and place it in a part of the garden where it won't be touched by people or pets.

Prickly pears have always been and continue to be a source of food and drink in Mexico and the southwestern United States. The young "petals," leaf pads, and fruits (called *tunas* in Spanish) are edible, though great care should be taken when harvesting. The spines and the smaller glochids are barbed and very difficult to remove from the skin. One American Indian tribe actually used them as fish hooks. The most harvested species of prickly pear are *Opuntia ficus-indica* in Mexico and *O. engelmannii* in the United States.

The suggested method of removing the spines is to harvest the fruit with long tongs, put the fruit in a pail of water, turn a high-pressure hose on them, and wash until the spines are dislodged. The spines float and can be discarded with the rinse water. Repeat this process until all the spines are removed. Extract the juice through a colander, then strain through cheesecloth.

The fruit was commonly eaten and could be dried for winter use. The flesh of the leaf pads or joints was used by the Cheyenne to thicken soups or stews. The Flathead made a poultice from the stems to treat backaches and infections. The Crow and the Navajo used the fruit as a red dye or as a mordant to fix dyes and colors.

The cochineal bug feasts on the prickly pear cactus and is the source, still, of red food coloring. The small black and white bugs suck on the sap and, as a means of deterring predators, produce carminic acid that is found in their gut. When this (along with the bug!) is ground into a powder and mixed with water, it turns bright red and is essentially tasteless, which makes it a good source for red food dye. The only commercially available alternative red dye is a synthetic petroleum product.

cochineal bug

CALIFORNIA POPPY

Eschscholzia californica

MILDLY TOXIC

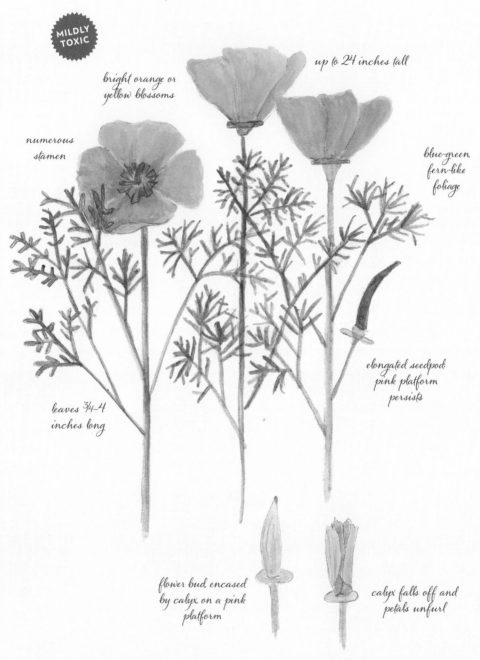

bright orange or yellow blossoms

up to 24 inches tall

numerous stamen

blue-green fern-like foliage

elongated seedpod: pink platform persists

leaves ¾–4 inches long

flower bud encased by calyx on a pink platform

calyx falls off and petals unfurl

BLOOMS: February–September.

HABITAT AND RANGE: Coastal areas, valleys, grasslands, savannas, hillsides, and open areas. Vancouver Island to Southern California below 7,000 feet. Naturalized elsewhere.

CONSERVATION: California poppy has become an invasive weed in Chile and Australia. The Joshua tree poppy (*Eschscholzia androuxii*), native to the area surrounding Joshua Tree National Park, is endangered.

WILDLIFE PARTNERS: Poisonous to livestock, avoided by most mammals and insects except blister beetles, which are attracted by the slightly spicy scent.

IN THE GARDEN: It is easy to grow from seed and is now planted almost throughout the United States. In warm areas, the plants go dormant during the winter; in cold areas, they are treated as annuals. Many cultivars have been developed with various colors. This makes a lousy cut flower as the petals fall off as soon as the flower is picked.

MEDICINAL USES: Be sure to consult a doctor before ingesting this plant as it is mildly toxic. The sap contains alkaloids. It was used by American Indian tribes as a mild sedative, a treatment for insomnia, and as pain relief for toothaches and headaches. It is sometimes now mixed with other herbs to treat anxiety, insomnia, agitation, and bedwetting. *Mildly toxic.*

There are eleven species of *Eschscholzia* native to the United States. *E. californica* is the California state flower.

Spanish settlers in California called it *dormidera,* meaning "the drowsy one," referring to its sedative properties. They believed that if you put the blossoms under a child's bed, the child would sleep better. They used the yellow pollen as a dye and as a cosmetic to make eye shadow and body paint. American Indians mixed the California poppy with bear fat and used it as a hair tonic to make the hair thick and glossy. Another Spanish name was *copa del ora* or

"cup of gold," and legend held that the plant filled the soil with gold.

This plant was found on the California coast in 1816 by a naturalist on board the Russian exploration ship, *Rurik*. He named the plant for the ship's surgeon, Johann von Eschscholtz.

American Indian women believed that the plant was a strong aphrodisiac, but its use was forbidden by many tribes, and if a woman was caught using it, she could be expelled from the tribe.

The seeds are in the plant's long, slender seedpods. When they burst open, they make an audible popping sound. The seeds are expelled so fiercely they can travel up to 6 feet.

CATTAIL
Typha latifolia

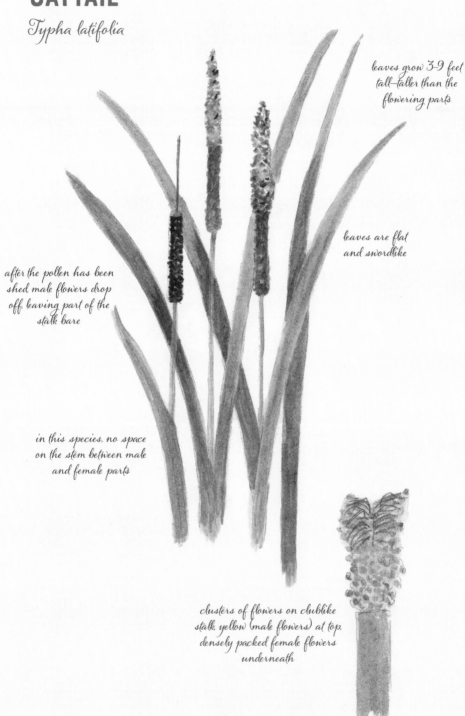

leaves grow 3-9 feet tall–taller than the flowering parts

leaves are flat and swordlike

after the pollen has been shed male flowers drop off, leaving part of the stalk bare

in this species, no space on the stem between male and female parts

clusters of flowers on clublike stalk, yellow (male flowers) at top, densely packed female flowers underneath

BLOOMS: May–July.

HABITAT AND RANGE: This plant is native to all states (except Hawaii) in fresh water, marshy areas, and sometimes in slightly brackish waters.

CONSERVATION: Cattail is considered invasive in some areas and often outcompetes other native wetland species, reducing diversity.

WILDLIFE PARTNERS: This plant is an important food and cover plant for wildlife. The cattail is so essential to pheasants that the number of cattails determines the number of pheasants that will winter in a particular area. In addition, it attracts red-winged blackbirds and the foliage is used by muskrats in building their lodges. It provides a landing site for dragonflies and damselflies.

MEDICINAL USES: In Africa, the root of cattail was given to women (and animals) as an aid in childbirth. American Indian tribes boiled it in milk to treat diarrhea.

Cattail is most conspicuous in early spring when seeds develop downy parachutes. The fluffy material was used for stuffing in quilts and pillows. When put in boots or shoes, it provided some insulation from the cold.

Many American Indian tribes wove cattail leaves into mats and rush seats, and the flowering stalk was dipped in fat or tallow to use as torches. Cosmetic dusting powder is still made from collecting pollen from the male flowers, and the female flowers are used for tinder.

American Indians and early Western settlers ate the young shoots like asparagus. Immature flower stalks were boiled and eaten like corn on the cob. The root was roasted and ground into flour, and its pollen was added to bread dough or cereal. It is still considered a tasty treat to eat from the wild.

COREOPSIS, PLAINS

Coreopsis tinctoria

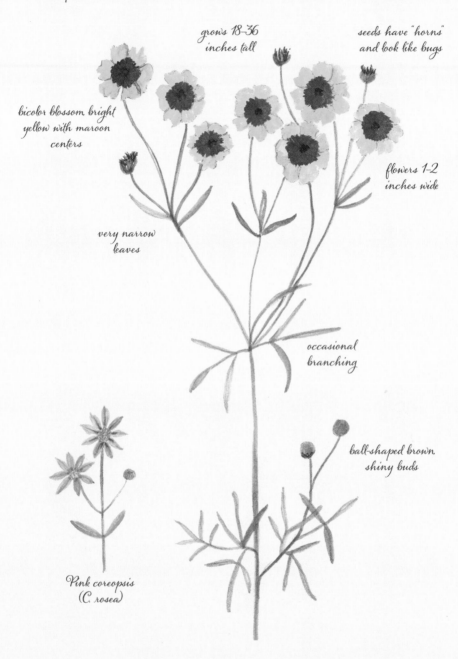

grows 18-36 inches tall

seeds have "horns" and look like bugs

bicolor blossom bright yellow with maroon centers

very narrow leaves

flowers 1-2 inches wide

occasional branching

ball-shaped brown shiny buds

Pink coreopsis (C. rosea)

BLOOMS: June–September.

HABITAT AND RANGE: Plains coreopsis (*C. tinctoria*) is an annual native to the eastern United States, but it is now naturalized and hardy in all states except Nevada and Utah. It prefers low, moist ground, roadsides, and open fields.

CONSERVATION: Pink coreopsis (*C. rosea*) is threatened and is protected in most areas of its natural range.

WILDLIFE PARTNERS: This plant is host to 42 different pollinators. Finch and sparrows eat the seeds.

IN THE GARDEN: Coreopsis is excellent for a sunny garden. It reseeds readily. Among the best cultivars is 'Summer Sunshine' developed from swamp tickseed (*C. palustris*). It is heat, drought, and deer resistant once established.

RELATED SPECIES: Lanceleaf coreopsis (*Coreopsis lanceolata*) is a beautiful perennial wildflower and a great addition to the garden. The bright yellow ray flower petals are notched at the ends, and the disc flower petals at the center are dark.

MEDICINAL USES: Many American Indian tribes made a tea from the root for stomach aches and diarrhea.

Coreopsis was first described by the Lewis and Clark expedition in 1805.

The genus name comes from two Greek words, *koris* (bedbug) and *opsis* (resembles or looks like), because the seeds were thought to look like bedbugs. Western settlers used dried plants to stuff mattresses and repel bugs. The name tinctoria means "of the dyers." The flowers of this species were used by settlers to make a golden, reddish dye. Other common names include tickseed, golden wave, calliopsis, and dyer's tickseed.

According to superstition, tea made from this plant was thought to protect one against lightning.

The Zuni believed that pregnant women who drank coreopsis tea were likely to have girl babies.

According to the language of flowers, coreopsis means "always cheerful." All species make excellent cut flowers.

This is the state wildflower for Florida.

EVENING PRIMROSE

Oenothera biennis

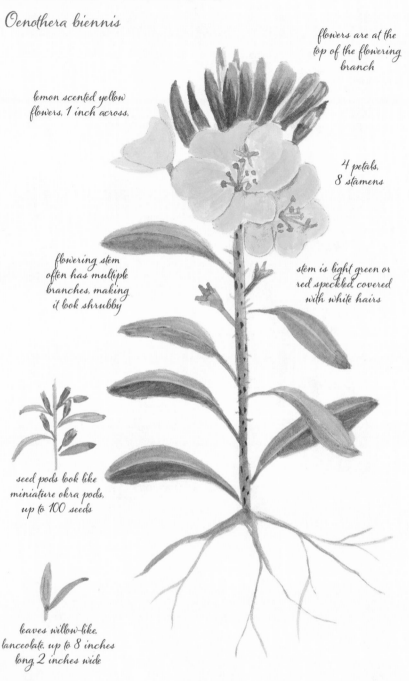

flowers are at the top of the flowering branch

lemon scented yellow flowers, 1 inch across,

4 petals, 8 stamens

flowering stem often has multiple branches, making it look shrubby

stem is light green or red speckled covered with white hairs

seed pods look like miniature okra pods, up to 100 seeds

leaves willow-like, lanceolate, up to 8 inches long, 2 inches wide

BLOOMS: Late spring–late summer.

HABITAT AND RANGE: Evening primrose (*O. biennis*) is found in sunny, arid, often disturbed places from Newfoundland south to Florida, west to Texas, and north to Alberta. Found sporadically in the West.

CONSERVATION: *O. biennis* and *O. speciosa* are both common and can be aggressive growers. Other species, such as wolf's evening primrose (*O. wolfii*), which grows in coastal areas of the Pacific Northwest, is a "plant of conservation concern." The Antioch Dunes evening primrose (*O. deltoides*) of California and the Colorado butterfly plant (*O. coloradensis*) are also of concern.

WILDLIFE PARTNERS: Hummingbirds, honeybees and bumblebees, and several moth species, including the primrose moth (*Schinia florida*), eat the nectar. Several caterpillars eat the foliage, including those of the pearly wood nymph, grape leaf folder moth, and white-lined sphinx. Seeds are eaten by several bird species, including goldfinch.

IN THE GARDEN: This is a biennial plant that dies after flowering during the second year. The first year the plant forms a rosette low to the ground, but during the second year, the stem grows quickly, sometimes up to a height of almost 7 feet, though usually much shorter. The seeds are tiny and airborne and can stay viable in the ground for up to 70 years.

RELATED SPECIES: Pink evening primrose (*O. speciosa*) is sprawling and grows only ½ to 2 feet tall on thin, weak stems. It is native to the American Midwest and Southwest and is naturalized throughout the lower 48 states.

MEDICINAL USES: Both the Cherokee and the Iroquois made a tea from the leaves as a stimulant to treat "laziness and over fatness." The Iroquois chewed the roots and rubbed them on muscles with the hope of attaining strength. They also made a salve for skin ailments. The seeds contain an oil that is currently sold as an herbal supplement to aid in treating symptoms of rheumatoid arthritis, PMS, multiple sclerosis, and eczema.

Although the petals of O. *biennis* appear to be a solid color without veins, under UV light they show a distinct nectar guide pattern. The blossoms open rapidly at dusk and stay open until noon the next day, longer if the weather is cloudy.

The genus name, *Oenothera,* is Greek for "wine scenting." Presumably, the roots of plants from this genus were at one time used to flavor wines. All parts of O. *biennis* are edible and considered tasty:

ROOTS: raw or cooked like potatoes

LEAVES: eaten raw or cooked like spinach; made into tea

FLOWERING STEMS: young stems are peeled and eaten raw or fried

FLOWER BUDS: raw in salads, pickled, fried, or put in soups

FLOWERS: raw in salads and desserts or as a garnish

SEEDS: roasted (10 to 12 minutes at 350°F) and used like sesame seeds in salads or pastries; during World War II, the seeds were roasted and ground and used as a coffee substitute

GOLDENROD

Solidago species; illustration is <u>Solidago canadensis</u>

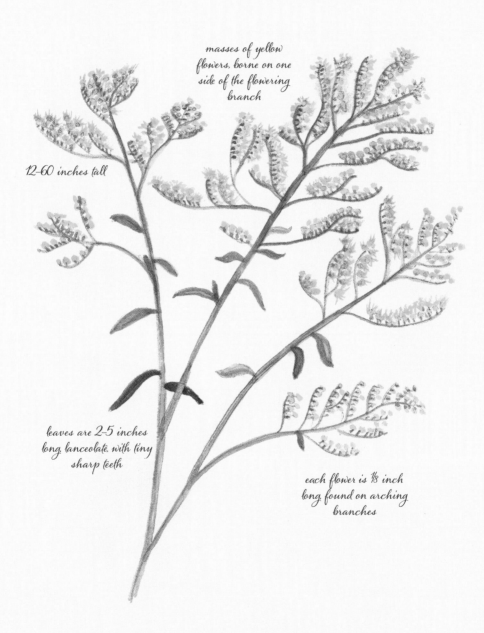

masses of yellow
flowers, borne on one
side of the flowering
branch

12–60 inches tall

leaves are 2–5 inches
long, lanceolate, with tiny
sharp teeth

each flower is ⅛ inch
long, found on arching
branches

GENERAL DESCRIPTION: There is tremendous variation among the 125 species of goldenrods native to the United States. Although identification to the species level is often difficult, it's relatively easy to tell when you're looking at a goldenrod. The flowers are tiny, daisylike, and have a distinctive bright yellow color. Many species, including seaside, rough-leaved, Canada, early, late, sweet, and rough-stemmed, have flowers on one side of the stem only. Leaves are generally entire, with variable serration along the edges.

BLOOMS: Goldenrods are short-day plants, blooming in late summer or early fall.

HABITAT AND RANGE: These plants grow throughout the United States. Though found most commonly in fields and roadsides, different species of goldenrods can be found in bogs or alpine ecosystems.

CONSERVATION: Some American goldenrod species are considered invasive and a threat to the biodiversity of European meadowlands. The result has caused reduced populations of European native grassland plants and an adverse effect on the insect and butterfly populations.

WILDLIFE PARTNERS: Pollinated by bees, flies, wasps, butterflies, and beetles, especially the goldenrod soldier beetle, also known as a Pennsylvania leatherwing. Goldenrods provide food for the larvae of many species in the *Lepidoptera* order of insects. The larvae sometimes form galls, which are often eaten by birds. Grouse eat the leaves. Goldfinch, junco, pine siskin, and sparrows eat the seeds. Beaver, porcupine, and rabbits eat the foliage. Deer will browse the entire plant.

MEDICINAL USES: Western settlers made a tea to treat inflammation, kidney stones, arthritis, toothaches, and sore throats. Salves and ointments (particularly from *Solidago virgaurea*) were used topically to treat sores, infections, wounds, and burns. The Okanagan used goldenrod to treat fevers.

In Europe, goldenrod was considered a powerful healing herb and was given the name *Solidago,* which means "to make whole" or "to heal."

American Indian healers included goldenrod when preparing a sweat lodge. They believed that this helped to remove all evil spirits from a person's body. The Navajo smoked the dried leaves, believing it would bring them good luck. The Meskwaki made a tea from the leaves to help a child who would not talk or laugh.

During the Revolutionary War, after colonists dumped imported tea into Boston Harbor, many plants, particularly goldenrod, were used to make a substitute tea called "liberty tea."

During World War II, sources for commercial rubber were scarce. Henry Ford spent a great deal of time and energy trying to develop an alternative source. For this quest, he turned to George Washington Carver, who experimented with many different kinds of plants, including goldenrod. His experiments resulted in a 12-foot-tall goldenrod plant that yielded a promising 12 percent rubber. Unfortunately, the more typical yield was only 7 percent and the product was of a poor quality, making it unfit as a viable source of rubber.

Contrary to the commonly held belief that goldenrod causes hay fever, it does not. The culprit is ragweed, which blooms at the same time.

Goldenrod is the state flower of Kentucky, Nebraska, South Carolina, and Delaware.

JACK-IN-THE-PULPIT

Arisaema triphyllum

TOXIC

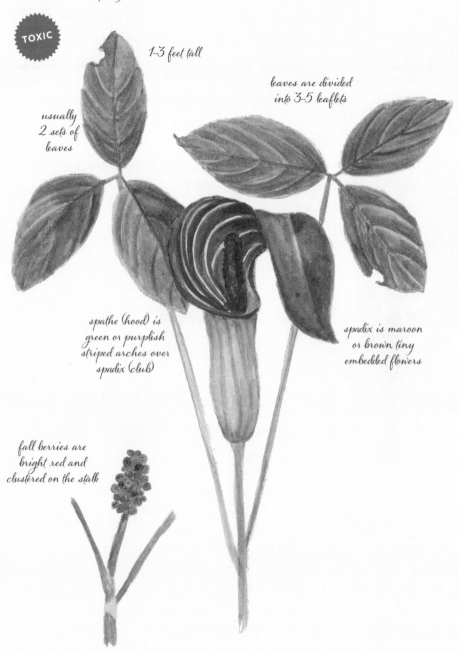

1-3 feet tall

leaves are divided
into 3-5 leaflets

usually
2 sets of
leaves

spathe (hood) is
green or purplish
striped arches over
spadix (club)

spadix is maroon
or brown tiny
embedded flowers

fall berries are
bright red and
clustered on the stalk

BLOOMS: March–June.

HABITAT AND RANGE: Common in moist, rich woods throughout the East, west to North Dakota, and south to Texas.

CONSERVATION: This plant is common.

WILDLIFE PARTNERS: The toxicity of this plant makes it unpalatable to most wildlife, though ring-necked pheasants, wood thrush, and wild turkey eat the seeds. It is pollinated by small flies and fungus gnats.

IN THE GARDEN: Care should be taken if included in the garden, due to its toxic nature. It cultivates well, though, and will spread to form colonies. Provide this plant with rich, moist soils and dappled shade. It takes three years for seedlings to bloom, but the plant can live as long as 25 years.

MEDICINAL USES: The dried root is made into a salve and used externally to treat a headache. *Toxic.*

The "pulpit" of this plant is the arching spathe, while the "Jack" is the spadix in the lower plant. The flowers are unisexual. To avoid self-pollination, the male flowers mature and die before the female ones do. The fall berries are bright red and are clustered on the stalk. Other common names include Indian turnip, bog onion, and brown dragon.

The toxic nature of the plant comes from calcium oxalate crystals that become imbedded in the skin when the plant is touched or eaten. The root is edible, thus the common name of Indian turnip, but it must be dried and boiled before eaten. American Indian tribes used the powdered root as flour.

The Meskwaki chopped up the raw root and mixed it with meat to give to their enemies. The meat masked the taste of the root, resulting in pain and discomfort from the poisonous plant.

According to superstition, the seed of Jack-in-the-pulpit was thought to tell the future of someone ill. When dropped into water, if it whirled around four times, clockwise, it was thought the patient would recover. If it went counterclockwise, the patient would die.

JEWELWEED, SPOTTED TOUCH-ME-NOT

Impatiens capensis

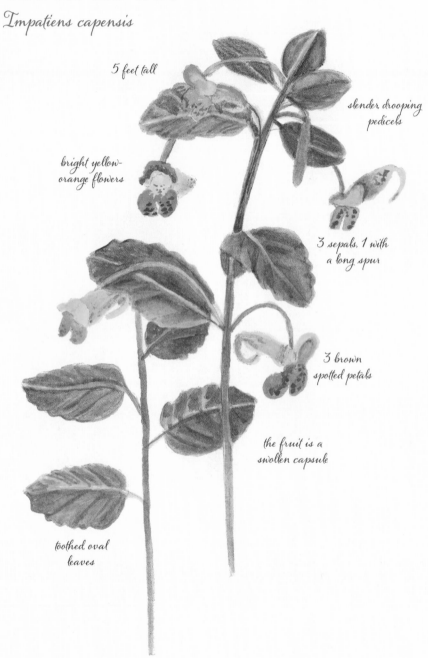

5 feet tall

slender drooping pedicels

bright yellow-orange flowers

3 sepals, 1 with a long spur

3 brown spotted petals

the fruit is a swollen capsule

toothed oval leaves

BLOOMS: July–frost.

HABITAT AND RANGE: Common in moist, shady places, especially along streams. Native from Newfoundland south to Florida, west to Texas, and north to North Dakota.

CONSERVATION: Very common; can be invasive.

WILDLIFE PARTNERS: This is a magnet for hummingbirds, butterflies, and bees and is considered an important pollinator plant. It is particularly attractive to the ruby-throated hummingbird and several species of long-tongued bees. Upland game birds such as the greater prairie chicken, bobwhite quail, and ruffed grouse eat the large seeds. White-footed mice also eat the seeds. White-tailed deer eat the foliage.

IN THE GARDEN: This is easily grown in moist, rich soils in part sun. It can become invasive, so be cautious. The cultivated impatiens from the same family come from East Africa and New Guinea.

MEDICINAL USES: Jewelweed contains chemicals with fungicidal properties. It has been used for centuries to cure skin ailments, such as athlete's foot, and to treat the rash resulting from poison ivy. Interestingly, jewelweed and poison ivy tend to grow close to one another.

The tightly coiled seedpods give this plant the common name, touch-me-not, since the slightest touch will cause it to explode and send seeds out a great distance. The name jewelweed comes from the fact that the plant repels water, and if a water droplet forms on the leaf, it shimmers like a jewel. Another possible reason for this name is that the flowers hang down like jewels on a necklace. American Indians called it crowing cock because of the shape of the flower.

Several American Indian tribes from the southeastern United States use jewelweed in medicine bags worn during the Green Corn Festival (see New York Ironweed, page 223), a ritual celebration lasting eight days that occurs in midsummer. Old fires are put out, both symbolically and literally, as debts and grudges are forgiven. It is considered a time of new beginnings and an opportunity to rekindle a sense of the sacredness of life.

LILY, GLACIER

Erythronium grandiflorum

6–15 inches tall

yellow (rarely white) flowers

1–5 nodding flowers on reddish-brown stems

6 reflexed petals, 6 stamens

anthers may be yellow, pink, red or purple

2 large (4–10 inch) leaves at base

elongated white corm grows deep

BLOOMS: May–July.

HABITAT AND RANGE: *E. grandiflorum* is often found in large colonies on moist slopes and shady areas at elevations between 3,000 and 9,500 feet. It grows from British Columbia and Alberta, south to California, and east to Colorado. It generally does not appear in Nevada.

CONSERVATION: The white form of glacier lily (*E. grandiflorum*) is threatened due to loss of habitat. The Minnesota dwarf trout lily (*E. propullans*) is on the US Endangered Species list.

WILDLIFE PARTNERS: Though difficult for humans to dig, apparently grizzly bears think it well worth the trouble. Studies indicate that bears will go out of their way to dig and eat these corms. Corms are also eaten by ground squirrels. The foliage provides the bulk of the diet of mule deer in May, when the new shoots first appear. It is pollinated by bumblebees and sometimes hummingbirds.

IN THE GARDEN: This is best enjoyed in the wild as it is reputedly difficult to grow in cultivation.

RELATED SPECIES: *Erythronium americanum*, dogtooth violet is very similar but is distinguished by mottled green and/or brown leaves. It is commonly found in spring in forests and open woods from eastern Canada to the northeastern United States, south to Georgia, and west to the Mississippi River. The raw plant, though not the root, was used as a contraceptive by American Indians. Other common names include dogtooth violet, yellow trout lily, yellow adder's tongue, and fawn lily because the two erect leaves look like a fawn's ears.

The genus name is from the Greek word for red, *eruthros,* probably in reference to a red-flowered species native to Europe. Other common names for glacier lily are avalanche lily, yellow fawn lily, dogtooth violet, and lamb's tongue. The latter names refer to the shape of the corm.

According to superstition, tea made from *Erythronium* species was thought to be effective in stopping hiccups. The corms were eaten and used by American Indian tribes in spite of the difficulty in digging them. The Okanagan considered it an important food source. Both corms and flower buds were eaten raw, boiled, baked, or made into pudding. Newly emerging leaves were eaten raw in salads or cooked like a potherb. The crushed corms were sometimes used as a poultice for boils.

SKUNK CABBAGE, AMERICAN

Lysichiton americanus

TOXIC

leaves
12–60 inches

flowers are tiny, densely
packed on stout stalk

leaves strongly
veined

bright yellow bract,
opens on one side,
8 inches long

flowers appear
before leaves grow
to their full length

BLOOMS: Late winter–early spring.

HABITAT AND RANGE: This is a wetland plant, preferring swamps and bogs. It is common in the Pacific Northwest and grows in Alaska, Canada, Idaho, Montana, Oregon, Washington, and Wyoming.

CONSERVATION: This has become an invasive weed in several European countries, including those of the British Isles and the Netherlands.

WILDLIFE PARTNERS: The malodorous plant attracts beetles, which pollinate it and use it as a mating site. Bears and elk eat the root.

IN THE GARDEN: A good addition to a bog garden, skunk cabbage grows slowly but surely and can become invasive if care is not taken. It grows in shade or sun and creates long-lasting populations of plants. It can be propagated by seeds or divisions.

MEDICINAL USES: Pacific Northwest Coastal tribes used the mashed root externally on burns, sores, and swelling. The sap from the raw root was used to treat ringworm. The heated blossoms were made into a poultice for rheumatism, and the heated leaves were used to help draw out thorns or splinters. *Toxic.*

Also called yellow skunk cabbage or swamp lanterns, this is a very prevalent plant in swampy areas of the Mountain West and Pacific Northwest. The huge leaves cover wide areas, and individual plants have been known to live 80 years or more. Western American Indian tribes  used the leaves to line baskets or as a wrap for smoking or roasting salmon or meat. The leaves were also used for collecting berries. It has the largest leaf of any North American native plant growing in its natural habitat.

According to superstition, the mashed root, applied to a child's head, was thought to make the hair grow more quickly. Smoke from burning the roots was thought to dispel bad dreams.

SKUNK CABBAGE, EASTERN

Symplocarpus foetidus

TOXIC

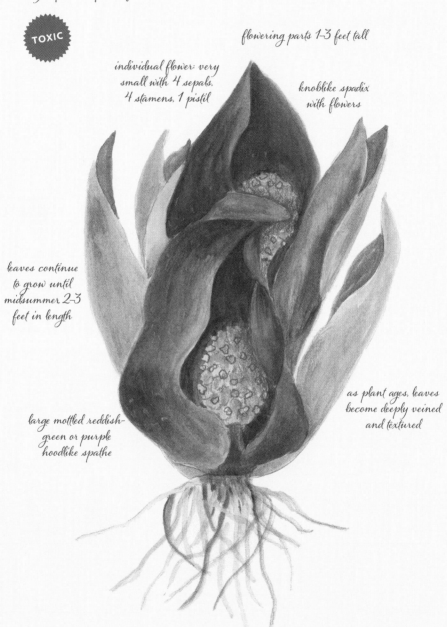

flowering parts 1–3 feet tall

individual flower: very small with 4 sepals, 4 stamens, 1 pistil

knoblike spadix with flowers

leaves continue to grow until midsummer 2–3 feet in length

as plant ages, leaves become deeply veined and textured

large mottled reddish-green or purple hoodlike spathe

BLOOMS: February–May.

HABITAT AND RANGE: This is a wetland plant, preferring swamps and marshes from Nova Scotia south to Tennessee and Georgia, and west to Minnesota and Iowa.

CONSERVATION: Though it is considered endangered in Tennessee, it is locally abundant in the more northern reaches of the range.

WILDLIFE PARTNERS: The musky scent of the flowers attracts flies and beetles.

IN THE GARDEN: Skunk cabbage is sometimes included in wetland gardens and is easily propagated by seed or rhizome segments sown in the fall.

MEDICINAL USES: American Indian tribes made a tea made from roots to treat venereal diseases. They also smelled the crushed leaves as a cure for a headache and made the raw root into a salve to relieve pain and swelling from rheumatism. Root hairs were used to alleviate pain from a toothache, and a tea made from these hairs was used externally to stop bleeding. This plant was listed in the *United States Pharmacopeia* from 1820–1882. It was used by the Western settlers for relief from spasms or cramps and to treat coughs, asthma, lockjaw, epilepsy, and rheumatism. *Toxic. Handle with care.*

All parts of the plant are poisonous except for the very young, still-curled leaves and the root, when prepared correctly.

Skunk cabbage (like American lotus, see page 96) is a thermoregulatory plant. That is, it generates heat, actually melting the snow as the plant emerges in late winter. This heat not only helps protect the bud from cold temperatures but also intensifies the odor of the blossoms and thus helps to draw pollinators.

The species name, *foetidus,* is Greek for "evil smelling." The common name, skunk cabbage also refers to the unpleasant odor. The genus name, *Symplocarpus,* is from two Greek words and means "connected fruit," and refers to the fruiting stalk, which comprises multiple ovaries growing together.

SUNFLOWER

Helianthus annuus

grows 2-13 feet tall

yellow ray flowers, red or brown disk flowers

flower heads are 3-6 inches across

flower heads are terminal on stems

leaves on petioles

alternate triangular leaves, entire or coarsely toothed

crossbreeds in the wild creating much variation

BLOOMS: June–September.

HABITAT AND RANGE: Originally native to the prairies, and now common throughout North America, it prefers sunny, open spaces such as fields, roadsides, and plains.

CONSERVATION: Common and abundant.

WILDLIFE PARTNERS: Many different bird species eat the seeds, including Wilson's snipes, doves, ruffed grouse, ring-necked pheasants, quail, blackbirds, bobolinks, lazuli buntings, horned lark, white-winged crossbills. Mammals eat the seeds, too, including mice, squirrels, gopher, lemmings, and prairie dogs. It provides nectar for many insects and butterflies.

IN THE GARDEN: Many cultivars have been selected for height, flower color and size, and abundant seed production. Among the best cultivars for the home garden include some that reach 12 to 14 feet in height. Called 'Kong' and 'Sunzilla,' these live up to their names with flower heads that can measure 24 inches or more across. Sunflowers need full sun but are drought tolerant once established.

MEDICINAL USES: The Cherokee used it for problems with their kidneys, the Southern Paiute used it for rheumatism, the Zuni for snakebite, and the Lakota for chest pain. Spanish explorers believed that it was an aphrodisiac.

The common name comes from the fact that the flower heads move to face the sun each day, pointing east in the morning and west in the afternoon. The genus name, *Helianthus,* comes from two Greek words, *helios* meaning "sun" and *anthos* meaning "flower."

Carved sunflower disks were found in prehistoric sites in Arizona, where it is thought that they were used in rituals and ceremonies. The Incas of Peru worshipped the sunflower as a symbol of the sun, and priestesses wore necklaces of sunflowers made of gold.

Sunflowers have been domesticated by indigenous peoples for 3,000 years. It is thought that the original sunflower seeds were quite small, but the size increased 1,000 times over through selective breeding. The seeds were eaten raw or roasted or pressed for oil. Flower buds were boiled or steamed and eaten, and the tender new stalks were eaten like asparagus.

Western settlers planted sunflowers near their homes in the belief that they would help protect against malaria.

The Hopi believed that when sunflowers were abundant, there would be a good harvest of corn. They often wore sunflowers in their hair during various tribal ceremonies. The Lakota said that when sunflowers are tall and in full bloom, the buffaloes are fat and give the best meat. The Gros Ventre used sunflower seed oil to paint their bodies for ceremonies. The Hopi and other tribes used the seeds to make a dark dye to color baskets. A yellow dye was created from the raw flowers.

This is the state flower of Kansas.

WHITE
FLOWERS

Anemone, wood *Anemone quinquefolia*

Bear grass *Xerophyllum tenax*

Bindweed, hedge (wild morning glory) *Calystegia sepium*

Black cohosh *Actaea racemosa*

Bloodroot *Sanguinaria canadensis*

Bunchberry *Cornus canadensis*

Corn lily (false hellebore) *Veratrum californicum*

Cow parsnip *Heracleum maximum*

Ginseng, American *Panax quinquefolium*

Goldenseal *Hydrastis canadensis*

Indian hemp *Apocynum cannabinum*

Indian pipe *Monotropa uniflora*

Lotus, American *Nelumbo lutea*

Mayapple *Podophyllum peltatum*

Onion, nodding *Allium cernuum*

Orchids *Goodyera pubescens, Platanthera psycodes, Platanthera dilatata, Calypso bulbosa, Spiranthes romanzoffiana, Epipactus giganteum, Platanthera ciliaris*

Partridgeberry *Mitchella repens*

Pipsissewa, common *Chimaphila umbellata*

Pokeweed *Phytolacca americana*

Sego lily *Calochortus nuttalii*

Snakeroot, white *Agertina altissima*

Solomon's seal *Polygonatum biflorum*

Strawberry, wild *Fragaria virginiana*

Water lily, American *Nymphaea odorata*

Yarrow *Achillea millefolium*

Yucca *Yucca filamentosa*

ANEMONE, WOOD

Anemone quinquefolia

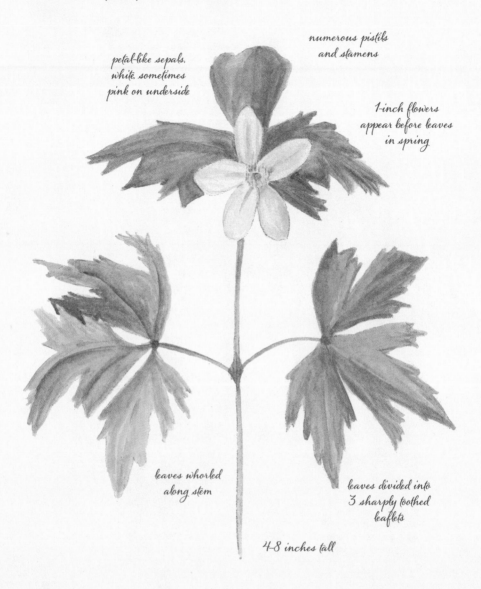

petal-like sepals, white, sometimes pink on underside

numerous pistils and stamens

1-inch flowers appear before leaves in spring

leaves whorled along stem

leaves divided into 3 sharply toothed leaflets

4-8 inches tall

BLOOMS: April–June.

HABITAT AND RANGE: Wood anemone (*A. quinquefolia*) is common in woods, thickets, and clearings throughout the East, west to the Dakotas, south to Mississippi and Arkansas, and east to Georgia.

CONSERVATION: This plant is abundant and secure. Carolina anemone (*A. caroliniana*) is considered endangered in Tennessee and Wisconsin.

WILDLIFE PARTNERS: Grouse and wild turkey eat the leaves and seeds. Sparrows eat the seeds. Rabbits, mice, and chipmunks eat the leaves. Pronghorn, deer, and moose eat the entire plant.

IN THE GARDEN: This is a good flower to include in a woodland garden or rock garden as it withstands a variety of conditions. However, it requires patience as it sometimes takes the plant five years to bloom.

RELATED SPECIES: In the West, blue anemone (*Anemone oregana*) is found in Washington and Oregon. It has 5 to 8 bluish sepals and between 35 to 100 stamens. *A. drummondii*, a low growing western anemone, grows in mountainous regions from Alaska to California.

MEDICINAL USES: The Meskwaki put the leaves on burns, used the root for headaches, and made a tincture from roots as a general tonic for "crazy people." The Ojibwe made a poultice from the root for back pain and a root tea to clear the throat for better singing. The Chippewas used Canada anemone (*A. canadensis*) root to make a poultice for wounds and sores, and they used the leaves to treat nosebleeds.

Legend tells us that the original home of this flower was Mount Olympus, where the prevailing winds blow. The word for the genus *Anemone* comes from *Anemoi,* the composite name for four Greek wind gods. Pliny wrote that the plant won't bloom until the winds blow.

Another legend suggests that anemone was sacred to Venus, who cried for her lost lover, Adonis. Wherever her tears fell, anemones began to grow.

Shakespeare gave the anemone magical powers to create love. In *A Midsummer Night's Dream*, Oberon tells Puck to place anemone blossoms on Titania's eyes so that she would fall in love with the first thing she saw on awakening.

Picking anemones in spring was a Roman ritual. People said prayers while picking the flower, believing that this kept loved ones healthy for the coming year.

In Egypt, anemones were thought to spread sickness, and people held their breaths as they ran past the flowers. The Chinese called it the death plant and placed it on family graves.

Other common names include windflower, nimbleweed, mayflower, and woodflower.

The Ponca tribe of Nebraska used the wooly fruit of *A. cylindrica* as a good luck charm when playing cards. In Alaska, *A. narcissiflora* leaves were beaten with oil until creamy and then frozen to make ice cream.

BEAR GRASS

Xerophyllum tenax

3-6 feet tall

flowers in
terminal raceme,
usually elongated
sometimes rounded

long spiny hairs
along stem

leaves up to
3 feet long, turn
white with age

long narrow
grasslike leaves
at base of stem

BLOOMS: May–August, depending on elevation.

HABITAT AND RANGE: British Columbia south to California and east to Wyoming. Found in subalpine meadows and open woods in coastal ranges and subalpine regions of the Rocky Mountains.

CONSERVATION: Bear grass is abundant and secure. A similar species native to the Appalachian Mountains, eastern turkeybeard (*Xerophyllum asphodeloides*), is considered endangered or threatened throughout its range except in Virginia and New Jersey.

WILDLIFE PARTNERS: Several mammals eat the evergreen leaves in winter, including Rocky Mountain goats and deer. The fleshy part of the leaves, at the base, are eaten by bears, mice, pocket gophers, and others. Deer and elk eat the flowering stalks. These flowers produce no nectar, but it is pollinated by bees, flies, and beetles. Flowers are slightly fragrant, though the scent differs in different locations, perhaps to attract local pollinators.

IN THE GARDEN: Do not transplant from the wild. Fresh seeds should be sown outside in late summer and kept moist until they germinate and are established. Bear grass needs full sun and well-drained soils. Divide and replant in spring. This plant doesn't bloom reliably, but the leaves make a nice textural addition.

MEDICINAL USES: The root makes blood vessels constrict, and a poultice is good for cuts, wounds, and sprains.

The name bear grass comes from the fact that bears use the long leaves in their winter dens. Other common names include turkeybeard, bear lily, pine lily, elk grass, soap grass (because if you rub root pieces together, they produce a lather), quip-quip, squaw grass, and Indian basket grass. The latter names come from the fact that bear grass leaves were important to

American Indian weavers. The fibrous leaves could be woven tightly enough to make waterproof baskets. The braided  leaves were used as decorations on buckskin dresses and jackets and on jewelry. The older leaves, which turn white, were easily dyed.

Bear grass survives and even thrives with periodic burns, and it is one of the first plants to sprout after a forest fire. Before flowering, which is generally thought to occur every five to seven years, the rhizome puts out new growth. After flowering, the parent plant dies, but because the young offsets have already become established, the population persists.

Today, the leaves are gathered for the commercial florist industry.

BINDWEED, HEDGE (WILD MORNING GLORY)

Calystegia sepium

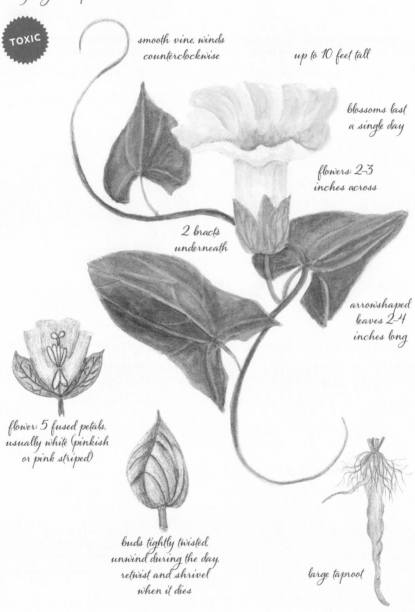

TOXIC

smooth vine, winds counterclockwise

up to 10 feet tall

blossoms last a single day

flowers: 2–3 inches across

2 bracts underneath

arrowshaped leaves 2–4 inches long

flower: 5 fused petals, usually white (pinkish or pink striped)

buds tightly twisted unwind during the day, retwist and shrivel when it dies

large taproot

BLOOMS: May–September.

HABITAT AND RANGE: Grows prolifically in almost all the lower 48 states on roadsides and cropland, empty urban lots, and along stream banks.

CONSERVATION: Considered a noxious weed in 46 states.

WILDLIFE PARTNERS: It is pollinated by long-tongued bees, including the morning glory bee, tortoise beetles, and many other types of insects. It is particularly useful to hummingbirds as a late summer to early fall bloomer. Bobwhite quail and ring-necked pheasants have been known to eat the seeds, though it is not an important food source for them.

IN THE GARDEN: Though *Calystegia sepium* is much too invasive to include in the garden, there are species in this family that are wonderful for the garden, such as the wild blue morning glory (*Ipomoea tricolor*).

RELATED SPECIES AND MEDICINAL USES: A fascinating species is the bush morning glory (*I. leptophylla*), native to the Great Plains regions. This plant can grow 2½ feet tall and more than 5 feet wide and develop a tap root 4 feet deep and more than 2 feet thick. The root, called a man root, served many different uses for the Lakota, including as a receptacle to store and carry embers. Many American Indian tribes used this medicinally. Raw, it was used to treat stomach aches. When smoked, it was thought to get rid of bad dreams. Ground into a powder, it was used as a stimulant for the heart and to revive people who fainted. The dried root was also used to help colts grow faster. It appeared in the *United States Pharmacopeia* from 1820–1863. The Chinese used bindweed root as a laxative, and in Mexico, it was used for both medicine and religious ceremonies. *Toxic.*

Bindweed is one of over 1,000 species found in the *Convolvulus* (morning glory) family. Though there is tremendous variation in the sizes and growth habits of all these species, all the flowers display a similar form of fused petals with a deep throat.

The twining, twisting growth habit of bindweed has given rise to many different common names, including devil's-gut, bearbind, and old man's nightcap.

Members of this family have been known and used in Japan since the 9th century when it was introduced from China and Korea. The Japanese love morning glories, and they use them to cover lattices for shade in the summers.

The Aztecs mixed the sap from the plant with rubber from the rubber tree to make bouncy balls.

Witches believed that these plants held special magic, especially when collected three days before the full moon and were useful for casting spells.

Morning glories growing in English gardens have earned the name "life of man," because it is a bud in the morning, opens fully by noon, and wilts by evening. In Victorian times, morning glories became part of the language of flowers, with blue symbolizing affection and the red blossoms meaning passion.

"I hold no preference among flowers, so long as they are wild, free, spontaneous."
—Edward Abbey

BLACK COHOSH

Actaea racemosa (formerly known as Cimicifuga racemosa)

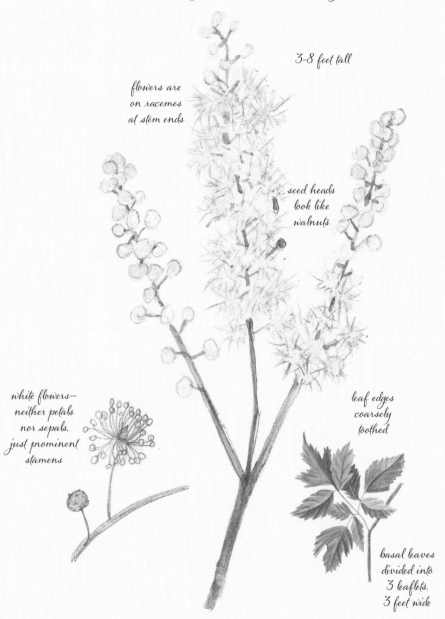

3–8 feet tall

flowers are on racemes at stem ends

seed heads look like walnuts

white flowers—neither petals nor sepals, just prominent stamens

leaf edges coarsely toothed

basal leaves divided into 3 leaflets, 3 feet wide

BLOOMS: Summer—earlier in the south, later in the north.

HABITAT AND RANGE: Found in woodland openings and prefers semi-shade, cool spots. The farther south the plant is found, the more shade it needs. Its natural range extends from southern Ontario, south to central Georgia, and west to Missouri and Arkansas.

CONSERVATION: Though locally abundant in most of its range, it is listed as rare in some states and appears on the endangered species list in Illinois and Massachusetts.

WILDLIFE PARTNERS: Stamens emit a sweet, fetid odor that attracts many insects, including flies, gnats, and beetles. It is a host plant for spring azure, holly blue, and Appalachian azure butterflies, and it provides nectar for butterflies such as the red admiral.

IN THE GARDEN: Several cultivars have been developed, including one with burgundy leaves. It needs part shade and rich, moist soils. It is a good specimen to use against a dark woodsy backdrop.

MEDICINAL USES: American Indian tribes used this plant to treat gynecological disorders, rheumatism, malaria, and as an aid in childbirth. European settlers used it for treating symptoms of menopause. In the 1950s, it was popular in Europe as an aid for women's health.

Black cohosh seems to have effects similar to that of estrogen, and it has been said to relieve some menopausal symptoms with moderate success. While more research needs to be done to determine its real effectiveness, it is currently a very popular herbal supplement. All black cohosh is wild collected, creating some concern for the future of its conservation.

Perhaps the most famous use of black cohosh was as an ingredient in Lydia Pinkham's Vegetable Compound, a general tonic sold from 1875 until the early 1900s. It was sold "For all those Painful Complaints and Weaknesses so common to our best female population . . . and is particularly adapted to the Change of Life."

The word *cohosh* is from the Algonquian word meaning "rough," and it refers to the dark, hard, knotted rhizome. The common name refers to the black or dark root. The plant has also been called fairy torches, because of its beautiful, graceful flowering stalks; rattle root or rattleweed because of the seed-pods; and bugbane because it was at one time used as an insect repellant.

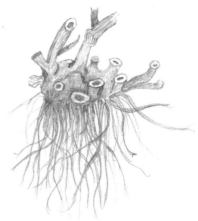

BLOODROOT

Sanguinaria canadensis

TOXIC

numerous petals,
bright golden
stamens

up to 12 inches tall

white blossoms,
1½-2 inches
across

palmate leaves
with 7-9 lobes

leaves continue
to grow until
midsummer 3-7
inches long

leaves stay curled
around stem until
blossom has been
pollinated

stem and root have
bright red sap

BLOOMS: Late February–April.

HABITAT AND RANGE: Found in woodland areas throughout the East, and west to the Mississippi River.

CONSERVATION: Bloodroot is abundant in most areas but is protected in New York and Rhode Island.

WILDLIFE PARTNERS: Bloodroot does not produce nectar, only pollen. It is pollinated by small bees and flies. The flowers are hermaphroditic, containing both male and female sexual parts. Its seeds are surrounded by a substance called elaiosome, which native ants eat. The ants then discard and disperse the intact seed.

IN THE GARDEN: A relatively easy wildflower to grow, bloodroot prefers a deciduous forest environment with sun in early spring and a bit of shade during summer months. It likes rich, loamy, and well-drained soils. Cultivars are available, displaying double flowers. One of the most beautiful is 'Multiplex,' which is the result of a genetic mutation causing all the stamens to transform into what look like a profusion of petals.

MEDICINAL USES: Teas and tinctures made from bloodroot were used both by American Indians and early Western settlers sparingly for colds and congestion, gastrointestinal, menstrual cramps, rheumatism, and fevers. European settlers used a drop of the sap in maple sugar to treat coughs. As a salve, it was used topically to treat wounds, burns, and infections. *Toxic. Do not ingest.*

All the various uses in folk medicine sparked a lot of interest in bloodroot. Scientists performed tests and found active alkaloids in the rhizome, indicating possible medicinal value. In recent times, bloodroot was included in toothpaste and mouthwash, but this was discontinued when it became apparent that it led to a premalignant condition. Bloodroot is toxic to humans and should not be ingested.

Algonquian-speaking tribes and the Iroquois called bloodroot *puccoon* or *poughkone,* a name given to several different plants that provided color and dye.

Both the genus name, *Sanguinaria,* which is taken from the Latin word for "blood," and the common name, bloodroot, were given to the plant because of the bloodred sap present in the stem and root. American Indians used this sap, mixed with bear grease or walnut oil, as war paint and a dye for blankets and baskets.

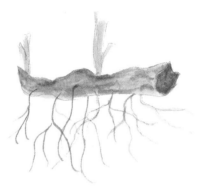

Today bloodroot is included in agricultural food supplements for livestock, particularly pigs. There is no indication of toxic side effects for these animals. Veterinarians have successfully used components from bloodroot to treat tumors in dogs and horses.

BUNCHBERRY

Cornus canadensis

8 inches tall

6 opposite leaves
(4 small 2 large)

flowers on
short stems
above leaves

4 white petal-like
bracts

true flowers only
1/16 inch across
with tiny petals

small mat, forming perennial slender
unbranched stems

underground rhizomes

leaves turn red in fall
(first veins, then entire leaf)

fruit is a drupe (like a cherry)
late summer or fall

flowers clustered together
in compound cyme,
surrounded by bracts

BLOOMS: Late spring–midsummer.

HABITAT AND RANGE: Found in large colonies in moist woodland areas throughout northern North America. It grows as far south as the Virginia Appalachian Mountains. It is also native to the mountains of Colorado and New Mexico.

CONSERVATION: Though abundant in most parts of its range, it is considered endangered in Illinois, Indiana, and Maryland, and it is threatened in Iowa and Ohio.

WILDLIFE PARTNERS: Bunchberry is pollinated by bees and flies. It is important forage food for caribou, moose, elk, and mule deer. The drupes (fruits) are eaten by song- and game birds, chipmunks and other small mammals, and black bears.

IN THE GARDEN: Bunchberry is a good plant for a woodland garden. It prefers cool, rich, moist soils and partial shade.

RELATED SPECIES: Flowering dogwood (*Cornus florida*) is a small understory tree found in deciduous forests of eastern North America. Western settlers used it to treat malaria, yellow fever, and dysentery. Small twigs of dogwood were used as toothbrushes by soldiers during the Civil War.

MEDICINAL USES: Similar in usage to the dogwood tree, bunchberry bark was made into a medicinal tea. The Abenaki took it for side pain, and the Lenape used it as an analgesic to ease pain. The roots were made into a tea for colicky infants.

This small woodland plant is one of the fastest moving plants on the planet! When the flowers are mature, the tiny petals become elastic and, when ready, spring back, releasing the filaments and anthers full of pollen. Pollen shoots out, accelerating at the rate of 24,000 meters per second! All of this takes less than a millisecond. To capture the action, you would need a camera that could take 10,000 frames per second. Because it explodes like a firecracker, it has earned the name crackerberry.

The fruits are edible, though tasteless. It was an important food source for Inuit on the North Bering Sea and the Arctic, who ate the fruit both raw and dried. The Hoh and Quileute of the Olympic Peninsula used the fruits in ceremonies. The people of the Nuxalk Nation in British Columbia and the Coast Salish on Vancouver Island smoked the leaves like tobacco.

In 2018, when Prince Harry of England married Meghan Markle, her veil was embroidered with floral images representing the 53 Commonwealth countries. Bunchberry was used to represent Canada.

CORN LILY (FALSE HELLEBORE)

Veratrum californicum

TOXIC

over 6 feet tall

tall, resembles
cultivated corn

flowers at top of
leafy, stout stalk

flowering stalk
branches

flowers white
or greenish
star shaped

leaves heavily
veined oval
12–15 inches

6 white tepals,
6 stamens,
3 branched pistils

BLOOMS: Midsummer.

HABITAT AND RANGE: Found in abundance in high mountain meadows at elevations from 3,500 to 11,000 feet near streams or in wet meadows. It is native from Alaska south to the Sierra Nevada mountain range, through Montana and Wyoming, and south to New Mexico and California.

CONSERVATION: Corn lily can be weedy and aggressive.

WILDLIFE PARTNERS: This plant is toxic and generally shunned by mammals. Even honeybees are harmed by the plant's toxicity, and the honey industry suffers in areas where this plant grows. It is larval food for the Setaceous Hebrew Character moth, so named because of the unusual markings on the wings of the adult, which look like the Hebrew character *nun*.

IN THE GARDEN: Because it is poisonous, it is not well suited for cultivation. A Eurasian native, black false hellebore (*Veratrum nigrum*), has beautiful dark purple flowers and is more often used in the garden.

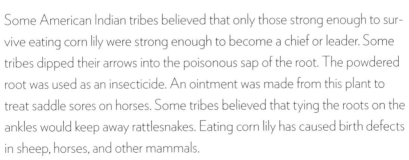

MEDICINAL USE: American Indian tribes used this for rheumatism, bruises, sores, wounds, cuts, snakebites, and burns. It was taken internally as a contraceptive, which was thought to result in permanent sterility and was used occasionally to treat venereal disease. *Toxic.*

Some American Indian tribes believed that only those strong enough to survive eating corn lily were strong enough to become a chief or leader. Some tribes dipped their arrows into the poisonous sap of the root. The powdered root was used as an insecticide. An ointment was made from this plant to treat saddle sores on horses. Some tribes believed that tying the roots on the ankles would keep away rattlesnakes. Eating corn lily has caused birth defects in sheep, horses, and other mammals.

This plant is also sometimes called false hellebore, though it bears little resemblance to true hellebores.

COW PARSNIP

Heracleum maximum

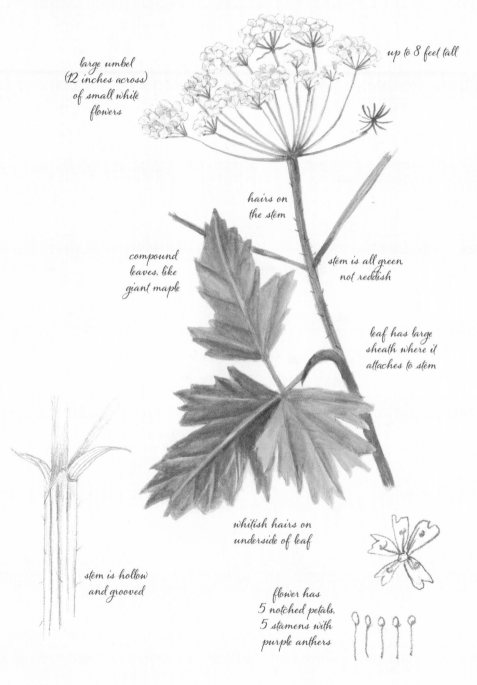

large umbel
(12 inches across)
of small white
flowers

up to 8 feet tall

hairs on
the stem

compound
leaves, like
giant maple

stem is all green
not reddish

leaf has large
sheath where it
attaches to stem

whitish hairs on
underside of leaf

stem is hollow
and grooved

flower has
5 notched petals,
5 stamens with
purple anthers

BLOOMS: April–September, depending on where it grows.

HABITAT AND RANGE: Grows throughout most of the United States from Southern California east to Georgia, north to Maine, and west to Alaska. It is scarce in the Southeast, except in Georgia. It prefers damp places such as meadows and stream banks.

CONSERVATION: Abundant in most of its range but endangered in Kentucky and threatened in Tennessee.

WILDLIFE PARTNERS: An important host plant for anise swallowtail butterfly. Forage food for deer, elk, moose, and bear. It is pollinated by bees, wasps, flies, and beetles.

IN THE GARDEN: As a biennial, this plant forms thick basal leaves the first year, attaining its final height in the second year. It needs full sun, a lot of space, and abundant moisture but makes a bold statement in a garden.

RELATED SPECIES: This plant looks like several others, including giant hogweed and water hemlock, which grows in similar habitats. Water hemlock is considered *the most poisonous plant* in America. Death from ingesting it can occur in 15 minutes, and for this reason, it was called "suicide plant" by the Iroquois. Though water hemlock looks much like cow parsnip, it's relatively easy to tell them apart. Water hemlock

Water hemlock

has a reddish stem and compound fernlike leaves, whereas cow parsnip has leaves that are rounded and lobed and an all-green stem. When in doubt, don't touch anything.

MEDICINAL USES: American Indian tribes made a tea from the root to treat intestinal disorders, headaches, and sore throats. A chewed wad was stuffed into dental cavities to relieve toothaches. As a poultice, it was placed on the skin to relieve pain from boils or skin disorders.

The genus was named for Hercules, both for its size and because he, reputedly, was the first to use it medicinally. It is also called Indian celery, pushki, and Eskimo celery.

American Indians gathered the large root and cooked it. They also ate the tender, young leaves, and they peeled and ate young shoots raw. Ashes from burned leaves were used as a salt substitute.

A piece of root carried as a good luck charm for hunters. The hollow stem was made into a whistle, used to call deer.

The hairs along the stem contain chemicals that, when exposed to sunlight, can cause a burning rash.

GINSENG, AMERICAN

Panax quinquefolium

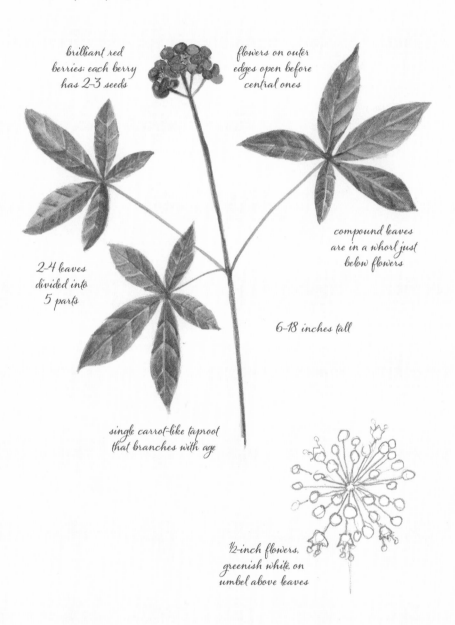

brilliant red berries: each berry has 2-3 seeds

flowers on outer edges open before central ones

compound leaves are in a whorl just below flowers

2-4 leaves divided into 5 parts

6-18 inches tall

single carrot-like taproot that branches with age

½-inch flowers, greenish white, on umbel above leaves

BLOOMS: June–July.

HABITAT AND RANGE: Rich, cool, moist woods from Ontario and Quebec south to Georgia, west to Louisiana and Oklahoma, and north to South Dakota and western Wisconsin. It is particularly prevalent in the Appalachian and Ozark regions.

CONSERVATION: Though at one time ginseng was common and widespread, overharvesting has taken its toll. The plant is now protected in 31 states and is considered endangered in at least two states. By law, harvested plants must have at least three leaves, and the ripe seeds must be replanted immediately.

WILDLIFE PARTNERS: Deer and small mammals eat the leaves. It is pollinated by the syrphid fly and halictid bees. The seeds are dispersed by the wood thrush and other songbirds.

MEDICINAL USES: In China, ginseng is prized as an aphrodisiac and a heart stimulant. American Indians used ginseng to treat coughs, headaches, and fevers and to strengthen thoughts. Today it is most often used as a general tonic to alleviate stress, promote relaxation, lower blood sugar, and boost energy.

Wherever it grows, ginseng has been revered for its medicinal value. The Cherokee name for this is "plant of life." It is the forked root that holds the magic, for it is thought to look "manlike" or "trouser-shaped," as the Chinese word *Jin-chen* describes it. In China, ginseng was considered a "dose of immortality." The genus name, *Panax*, comes from the Greek meaning "cure-all."

Both the size of the root and the number of leaves increase every year

Even as early as 1824, ginseng was an important export from the United States. In that year, 750,000 pounds of native ginseng were exported to China. Ginseng *can* be cultivated, though it is a difficult and time-consuming crop to grow. Ninety-five percent of the ginseng grown in the United States comes from Wisconsin. But cultivated ginseng is not as valuable nor as potent as wild collected, so there is still a huge market for wild-dug 'sang,' as the old-timers called the plant, making laws and regulations important for protecting this species. Only older plants, indicated by the number of leaves, can be dug legally.

GOLDENSEAL

Hydrastis canadensis

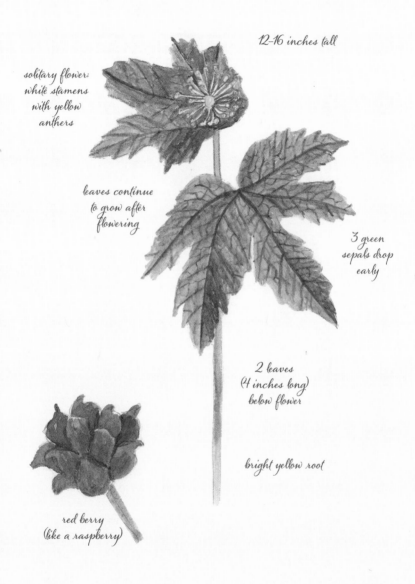

12-16 inches tall

solitary flower:
white stamens
with yellow
anthers

leaves continue
to grow after
flowering

3 green
sepals drop
early

2 leaves
(4 inches long)
below flower

bright yellow root

red berry
(like a raspberry)

BLOOMS: April–May.

HABITAT AND RANGE: Rich woods in Vermont south to Florida, west to Mississippi, and north to Iowa and Minnesota.

CONSERVATION: Considered threatened or endangered due to overharvesting in 12 eastern states.

WILDLIFE PARTNERS: Pollinated by bees and flies. Birds, such as wild turkey, and small mammals, such as moles and voles, eat the seeds.

IN THE GARDEN: Goldenseal can be propagated by seed in fall or spring, and it prefers a shady area with rich, moist soils.

MEDICINAL USE: American Indians used the yellow root as a laxative, tonic, astringent, and stimulant. Made into a salve, it was used as an antiseptic for skin infections and to help wounds heal. The powdered root was used to treat inflammation of the throat and eyes and mouth ulcers.

American Indian tribes used the bright yellow root as a dye and paint. Other common names include orangeroot, Indian turmeric, eyebalm, and ground raspberry, which is a reference to the single red berry that the plant produces. Lewis and Clark described the plant as being "a sovereign remedy for sore eyes."

Some tribes mixed the sap from the root with animal fat to use as an insect repellant.

INDIAN HEMP

Apocynum cannabinum

TOXIC

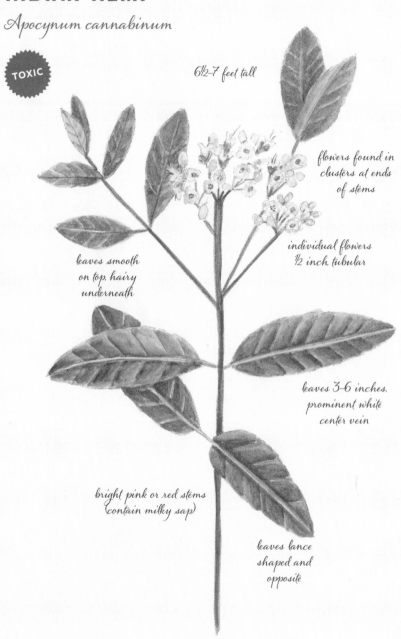

6½–7 feet tall

flowers found in
clusters at ends
of stems

individual flowers
½ inch tubular

leaves smooth
on top, hairy
underneath

leaves 3–6 inches,
prominent white
center vein

bright pink or red stems
(contain milky sap)

leaves lance
shaped and
opposite

BLOOMS: July–August.

HABITAT AND RANGE: Very common in shady areas, hillsides, and near streams. This plant is native throughout the United States.

CONSERVATION: Considered an aggressive weed in several states.

WILDLIFE PARTNERS: Host for the larvae of two species of hummingbird moths. It is pollinated by both moths and butterflies.

IN THE GARDEN: Not recommended for cultivation as it can easily become aggressively invasive.

MEDICINAL USE: Various American Indian tribes harvested the root in autumn, dried it, and made it into a tea for coughs, fevers, rheumatism, to increase lactation of nursing mothers, and as an aid to heart disease. *Toxic.*

There are several common names for this plant, including Indian hemp, dogbane, army root, wild cotton, and rheumatism root. The common name, dogbane (*bane* means to kill or get rid of), and the genus name *Apocynum,* which means "poisonous to dogs," refers to the poisonous qualities of the plant. The words *hemp* and *cotton* both refer to the common usage of the plant as a source of fiber.

American Indians used the inner fibers of the plant-like flax to make clothes, twine, cords, bags, nets, and fishing line. The fibers don't shrink or lose strength in water. Some tribes, particularly in the West and Northwest, used cordage made from Indian hemp to weave bags for holding everything from nuts and berries to clothing. A very strong thread made from Indian hemp was used to stitch pieces of leather together to make tepees.

The cords were also used by women to make "counting-the-days" balls or *ititamat.* When a woman married, she made one of these balls and used it to mark special days or events by tying knots or trinkets, shells, or beads to the string. She would then wind it into a ball and use it as a reminder of life events.

Men of various tribes used the hemp cords to tie eagle feathers to skirts worn in ceremonial dances.

INDIAN PIPE

Monotropa uniflora

MILDLY TOXIC

7-11 inches tall

leaves, also
translucent,
grow from stem
in sheaths

plant lacks chlorophyll
and is translucent,
appearing white

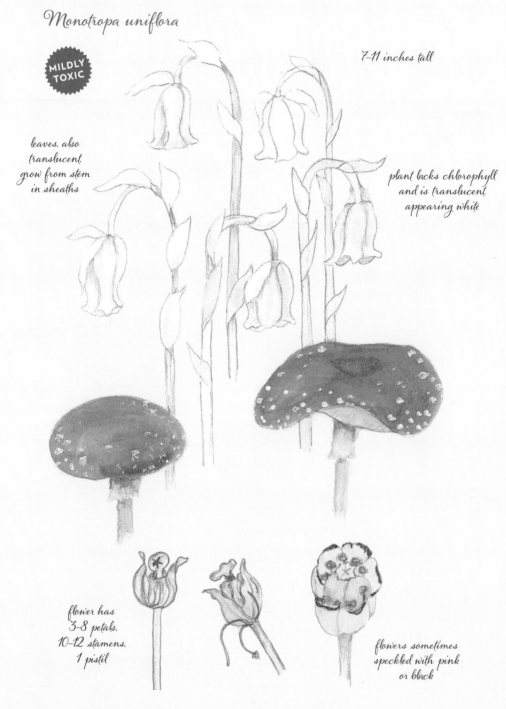

flower has
3-8 petals,
10-12 stamens,
1 pistil

flowers sometimes
speckled with pink
or black

BLOOMS: Early summer–fall.

HABITAT AND RANGE: This plant needs a lot of moisture but can grow in sun, part shade, or almost total shade. Because it depends on a fungus, which in turn depends on trees for nutrients, it is found in wooded areas. It grows in temperate regions throughout the world and is often associated with beech trees. In North America, it grows almost everywhere except a few southwestern and mountain states.

CONSERVATION: Though unusual, it is locally abundant within most of its range. It is considered endangered in California.

WILDLIFE PARTNERS: Bears eat both the flowers and the roots. It is pollinated by long-tongued bees.

IN THE GARDEN: Although it is parasitic, the plant is not a fungus but an actual flowering perennial. Because of its intricate dependency system, it is difficult, if not impossible, to propagate or move and cultivate.

MEDICINAL USES: The Cherokee used it as a tonic and antispasmodic. The mashed root was given to children who suffered from epilepsy or convulsions. The crushed flowers were used externally on warts and bunions. An eye wash was made from the sap of the plant. Flowers were used for toothaches. *Mildly toxic but is generally considered safe.*

Indian Pipe was one of Emily Dickinson's favorite flowers. She wrote of it:

> *White as an Indian Pipe*
> *Red as a Cardinal Flower*
> *Fabulous as a Moon at Noon*

An illustration of Indian Pipe was chosen for the cover of her first book of poetry.

Indian Pipe has several common names, including ghost plant, ghost pipe, and corpse plant. If this is picked, the plant turns black immediately. After the flower sets seeds, it turns brown or black and shrivels.

LOTUS, AMERICAN

Nelumbo lutea

receptacle has 10–20 pistils embedded on surface

flowers are pale yellow or white, 7–11 inches across

numerous golden yellow stamens hooked on ends

leaves and flowers held above water surface (from ½–3 feet)

leaves attached to central petiole

roots firmly rooted in mud

lotus seedpods

BLOOMS: Spring–summer.

HABITAT AND RANGE: Wetlands, swamps, ponds, and lakes from Minnesota south to Florida and west to Oklahoma.

CONSERVATION: Can be invasive in some areas.

WILDLIFE PARTNERS: This species is cross-pollinated by honeybees, native bees, and other insects. It is a larval host for the lotus borer and cattail borer moths. Canadian geese, mallards, and northern shovelers eat the seeds. Muskrats eat the rhizomes and leaf petioles.

IN THE GARDEN: This plant is a welcomed addition to a water garden. It is hardy north to Ontario and south to Florida and Texas. To bloom, it requires full sun and at least three months of temperatures between 75° to 85°F.

Worldwide, there are only two lotus species: the American lotus (*Nelumbo lutea*) and the sacred lotus (*N. nucifera*) of Asia.

The American species was an important food source for American Indian tribes. The large tuber, which can grow as big as a man's arm, was harvested in fall. The seeds were sometimes called alligator corn. American lotus was considered one of the "life symbol" plants of the Osage of Oklahoma. Called yonkapins, these plants were thought to be a sacred food and a symbol of health and long life. Both the Dakota and the Omaha believed it had mystical powers and used it in sacred ceremonies.

The seeds, which are 19 percent protein, can be popped like popcorn.

American lotus seeds stay viable for over 200 years. The seeds of the sacred lotus, *N. nucifera,* though, will last well over a thousand years.

The "lotus effect" was discovered by scientists who studied the remarkable capacity of the lotus leaf to shed water. When water hits the surface of the leaf, it immediately clumps into droplets, dragging any available dirt or mud with it. The water rolls off, leaving the leaf surface clean and dry. This property is known as ultrahydrophobicity.

Scientists testing *N. nucifera* at the Adelaide Botanic Gardens in Australia found that the plants maintained an even temperature between 86° and 95°F, even when outside temperatures dropped into the low 50s.

MAYAPPLE

Podophyllum peltatum

TOXIC

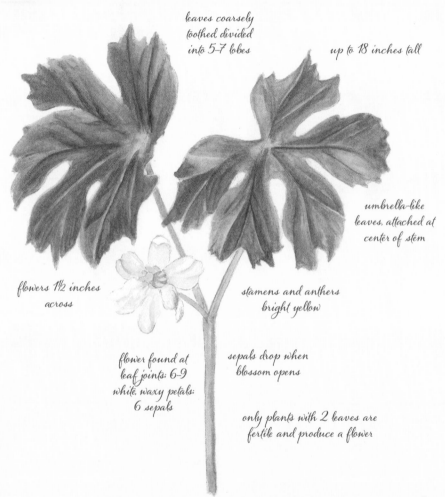

leaves coarsely
toothed divided
into 5–7 lobes

up to 18 inches tall

umbrella-like
leaves, attached at
center of stem

flowers 1½ inches
across

stamens and anthers
bright yellow

flower found at
leaf joints: 6–9
white, waxy petals;
6 sepals

sepals drop when
blossom opens

only plants with 2 leaves are
fertile and produce a flower

BLOOMS: April–June.

HABITAT AND RANGE: Prefers a rich, moist woodsy environment but is adaptable to varying conditions and can be found in open woods and shady roadsides. Grows throughout the East, from southern Canada, south to Florida, west to Texas, and north to Minnesota.

CONSERVATION: Abundant throughout most of its range, it is considered endangered in Florida. The only other species in the genus is found in the Himalayas, and it is on an international list of endangered species due to over-harvesting for medicinal use.

WILDLIFE PARTNERS: Mammals tend to shun the foliage because it is toxic and bitter. Bees pollinate the flowers. The larvae of the sawfly and some moth species eat the leaves. Several small animals eat the fruit, including the box turtle, possums, skunks, and raccoons.

IN THE GARDEN: Mayapple is easily propagated by planting rhizomes. It spreads quickly to form dense colonies. It can be slightly aggressive in some regions. The foliage dies back at the end of summer. It likes partial shade and is adaptable to varying soil conditions.

MEDICINAL USE: In spite of its toxicity, the plant was, and still is, used medicinally. American Indians used it as a purgative and as an insecticide. Early Western settlers used it to treat jaundice, liver, fever, syphilis, hearing loss, snakebite, parasitic worms, and cancer. Because chemicals within the plant are known to stop cell duplication, there was great hope that it would be effective in treating cancer. Today, the chemical podophyllotoxin is an FDA-approved drug and has been found to be somewhat effective in treating genital warts. *Toxic.*

The genus name, *Podophyllum,* comes from Greek and means "foot leaf." The species name, *peltatum,* means shield shape. Both names refer to the prominent leaf. Another name is umbrella leaf, again referring to the large leaf.

The common name, mayapple, refers to the fruit, which ripens in late spring. It is round, yellow-green, and the size of a small lemon.

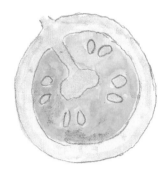

Other common names, such as wild lemon, hog apple, and raccoon-berry, also refer to the fruit, which is edible when fully ripe. Western settlers used it for making jams, jellies, and marmalades. A Southern drink was made by mixing mayapple juice, wine, and sugar.

An Appalachian superstition suggests that a girl who pulls up the root of the mayapple will soon become pregnant.

ONION, NODDING

Allium cernuum

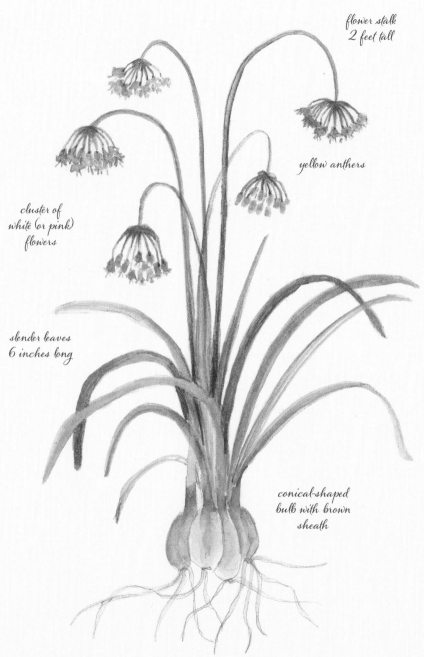

flower stalk
2 feet tall

yellow anthers

cluster of
white (or pink)
flowers

slender leaves
6 inches long

conical-shaped
bulb with brown
sheath

BLOOMS: July–August.

HABITAT AND RANGE: Dry woods, prairies, and rocky sites. The range is broad: east from New York and south to Florida, in the Ozarks, in the Rocky and Cascade mountain regions of the West, and from Mexico to Washington on the Pacific coast.

CONSERVATION: Most *Allium* species in North America are not at risk. However, the related wild leek (*A. tricoccum*) has been overharvested.

WILDLIFE PARTNERS: Bees are the primary pollinators, and nodding onions are especially important for honey bees. Because of the acrid taste of the leaves, the foliage is not generally eaten by mammals, but bears, squirrels, marmots, and others will dig the bulbs.

IN THE GARDEN: The showier wild onions, such as *A. cernuum*, are welcomed additions to a rock garden. Cultivars such as 'Major' display brighter colors and bigger flowers.

MEDICINAL USE: American Indian tribes used crushed leaves to alleviate the sting from bees and insects. The bulbs strung like a necklace were thought to keep you healthy.

Almost all regions of North America have some sort of native species of *Allium*. Perhaps the most famous of these is the wild leek (*A. tricoccum*), or ramp, as it is called in the South. Ramp festivals are extremely popular, especially in Appalachian regions. The bulbs are best gathered after the second year and are eaten raw, boiled, and pickled. Ramps have also been used to make salsa and jelly.

Early spring leaves were particularly welcomed as a fresh food by American Indian tribes and Western settlers. Native onions were sometimes eaten as a staple but were more often used as a condiment or for flavoring, as we use cultivated onions today.

Prairie onions, called skunk eggs due to their unpleasant odor, were baked into what the locals called SOB stew.

A. cernuum grew abundantly in the Great Lakes region. In the 1600s, the French explorer and fur trader La Salle reported that the Miami tribe felt this plant, which they called *checagou,* was so important that they named the Chicago River after it. It is also, perhaps, the origin of the name of the city.

People have been using onions for many millennia, and there is documented use of onions as far back as ancient Egypt, around 3200 B.C. Alexander the Great firmly believed that if you ate strong foods, you would become physically strong, and he insisted that his men eat onions to give them added strength for battle.

Onions get their odor and tear-inducing effect from volatile sulfuric gases that are released when its cells are bruised or cut.

ORCHIDS

Over 200 species of orchids are native to North America. They grow in every state. One hundred species are native to Florida, and 48 species are native to Minnesota. They display an astonishing variety of flower forms, colors, shapes, and sizes. All orchids have three sepals, three petals, and a floral column that houses the reproductive organs. The third petal often has a fantastic and strange shape. Leaves vary from large and thick to thin and narrow. Each orchid species has a complex and specific relationship with both a fungus and a pollinator. (See Pink Lady's Slipper, page 161). Most native orchids are difficult to establish in a wildflower garden because of this. They should never be dug from the wild, but many species are available through the nursery trade.

Orchids were named for an unfortunate youth, Orchis, the son of a nymph and a satyr. One evening, having drunk too much wine at a festival, he approached a priestess and grabbed her. The horrified crowd tore him apart, but his father begged the gods to make him whole again. The gods refused, but they did create the orchid from his body, bringing great beauty into the world.

ILLUSTRATED SPECIES

CONSERVATION: Many of the native orchids are included on the federal lists for species that are endangered, threatened, or of special concern. The greatest threat to native orchids is loss of habitat.

Rattlesnake plantain (*Goodyera pubescens*)
One of the most common of the native orchids in eastern regions, rattlesnake plantain is most easily identified by its beautiful evergreen leaves. The flowers are small, white, and arranged on a hairy stalk about 18 inches tall. There are between 20 to 80 evenly spaced tiny blossoms.

BLOOMS: July–August.

HABITAT AND RANGE: Common in both moist and dry woods from Nova Scotia south to Florida and west to Oklahoma and Minnesota. American Indians used it for a variety of medicinal purposes, and they also used it topically as a love potion. The common name comes from the vague resemblance of the leaves' markings to the markings on a rattlesnake.

Rattlesnake plantain

Lesser purple fringed

Lesser purple fringed (*Platanthera psycodes*)

Gorgeous, showy purple flowers with fringed petals. Flowering stalk up to 3 feet with up to 100 flowers. Similar to the yellow fringed orchid (*Platanthera ciliaris*).

BLOOMS: Late summer.

HABITAT AND RANGE: Native to northeastern, central, Great Lakes, and Appalachian regions of the United States.

CONSERVATION: It is threatened or endangered in several states, particularly in the southern part of its range.

Bog candle

Calypso orchid

Bog candle (*Platanthera dilatata*)

A tall, leafy stem holds many beautiful bilaterally symmetrical white flowers that look like little ghosts, torches, or candles. Grows 6 to 50 inches.

BLOOMS: Summer months.

HABITAT AND RANGE: Native to northern states across the United States, and south to New Mexico and Southern California in the West. Grows in boggy, wetland areas.

CONSERVATION: Widespread and abundant in most of its range

Calypso orchid (*Calypso bulbosa*)

The blossom grows 3 to 5 inches tall, has bright pink graceful petals, and a pouch with yellow hairs on the lower, central petal. It was named for Calypso, the nymph who waylaid Odysseus on his journey home. American Indians dug the corms for food. Tribes in British Columbia used it to treat epilepsy.

BLOOMS: Late spring–summer.

HABITAT AND RANGE: Native in many mountainous regions of the West, and to most of the northern states from Alaska, south to California, and east to Newfoundland. This orchid prefers moist areas and damp, coniferous woods.

CONSERVATION: This plant is considered endangered in many states.

Hooded ladies' tresses (*Spiranthes romanzoffiana*)

This orchid has a single stem with 3 to 6 basal leaves and up to 60 cream-colored flowers in 1 to 4 rows spiraling around the stem. It was named for the 18th-century Russian Count Nikolay Rumyantsev, who sponsored a North American plant exploration expedition.

BLOOMS: Summer–early fall.

HABITAT AND RANGE: Native to Alaska and most northern states, south to Southern California. It is found in moist, open spaces.

CONSERVATION: Locally abundant.

Hooded ladies' tresses

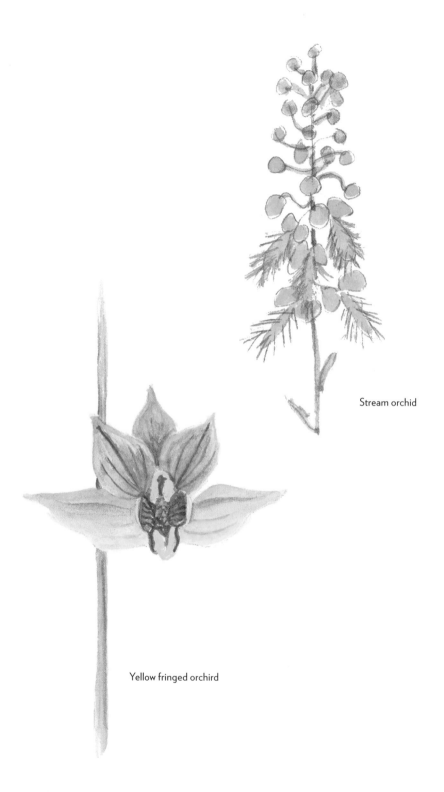

Stream orchid

Yellow fringed orchird

Stream orchid (*Epipactus gigantea*)

This plant has 5 to 25 greenish-yellow flowers in a loose raceme. Each blossom is striped with purple veins and has a bright yellow, purple, or orange lip. The blossoms have a fragrance like honeydew, attracting pollinators. When moved by the wind, the lower lip moves, as if the plant is talking or "chattering," thus the common name.

BLOOMS: Late spring–summer.

HABITAT AND RANGE: Also called chatterbox, this orchid prefers marshes, streams, floodplains, and shorelines, with no tolerance for dry or drought conditions. It is native throughout the West and is found east to Texas and South Dakota.

CONSERVATION: Considered threatened and of special concern.

Yellow fringed orchid (*Platanthera ciliaris*). Also seen as *Habenaria ciliaris*.

Yellow or orange-sherbet orange. The stem grows to 2 feet or more with leaves that "clasp" the stem. Upper leaves are small and narrow, while basal leaves can grow 10 inches or longer. The blossoms are found in a terminal cluster. Each individual flower has a deeply fringed lower lip, making this a stunningly beautiful flower. The Cherokee used the roots to make a hot tea to treat diarrhea. Taken cold, the tea relieved headaches. The Seminole used it for snakebites.

BLOOMS: Late summer–fall.

HABITAT AND RANGE: It grows in open woods and dry hillsides from southern New England, south to Georgia, west to Texas, and north to Michigan and Wisconsin.

CONSERVATION: This plant is extremely rare in New England and more common in the southern states.

PARTRIDGEBERRY

Mitchella repens

single red berry,
persists until next
blooming season

4 inches tall

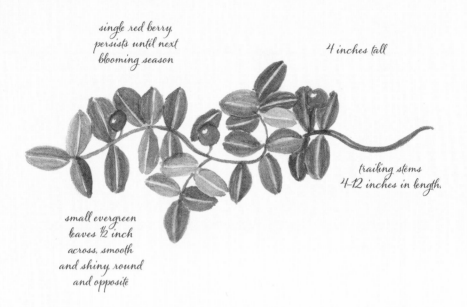

trailing stems
4–12 inches in length,

small evergreen
leaves ½ inch
across, smooth
and shiny, round
and opposite

each flower has
4 petals,
4 stamens,
1 pistil

small white (pinkish)
fringed blossoms in pairs

BLOOMS: May–July.

HABITAT AND RANGE: Common throughout eastern North America from Canada, south to Florida and Texas. It prefers shady conditions but withstands both damp and dry soils. Often found on slopes or streambanks.

CONSERVATION: Abundant.

WILDLIFE PARTNERS: This plant is cross-pollinated by bumblebees and is a source of nectar for many insects. Occasionally deer eat the foliage. Several species of birds, including grouse, quail, and wild turkey, eat the seeds. Small mammals such as the white-footed mouse, skunk, and fox also eat the berries.

IN THE GARDEN: Partridgeberry makes a good evergreen ground cover. It's a vine but does not climb and rarely becomes aggressive. Once established, it does well, particularly under rhododendrons and azaleas.

MEDICINAL USE: Cherokee women made a tea from the roots and drank it for weeks before delivery, believing that it would hasten childbirth and make labor easier. The tea was given to babies with a rash, newborns with upset stomachs, and adults with bladder or urinary trouble. The Menominee used a tea made from the leaves to treat insomnia and "diseases of women."

The genus was named for John Mitchell (1680–1768), who developed a method of treating yellow fever in 1745. The species name means "creeping." Common names include twinberry, squawberry, checkerberry, and running fox.

The flowers are dimorphous. Each flower of the pair is different from the other—one has a long pistil and short stamens, and the other has a short pistil and long stamens. This assures that the plant will not self-fertilize.

The berries are harmless but tasteless, and several different tribes mashed them to dry and make into cakes. The Ojibwe smoked the leaves during ceremonies.

PIPSISSEWA, COMMON

Chimaphila umbellata

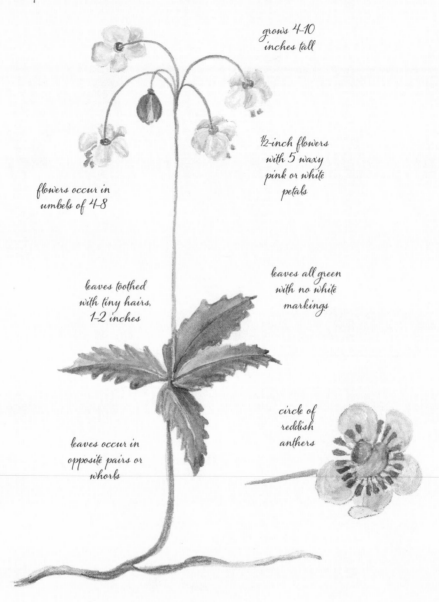

grows 4–10 inches tall

½-inch flowers with 5 waxy pink or white petals

flowers occur in umbels of 4–8

leaves toothed with tiny hairs, 1–2 inches

leaves all green with no white markings

leaves occur in opposite pairs or whorls

circle of reddish anthers

BLOOMS: Midsummer.

HABITAT AND RANGE: The common pipsissewa (*C. umbellata*) has a broad natural range and is native to all western and northeastern states. It likes dry woodlands and prefers well-drained, sandy soils.

CONSERVATION: Of special concern in some parts of its range. It is threatened or endangered in Iowa, Ohio, New York, and Illinois.

WILDLIFE PARTNERS: Pollinated by bumblebees, flies, and long-tongued bees.

IN THE GARDEN: Prefers dry, shady conditions. Needs very well-drained soils.

RELATED SPECIES: Spotted wintergreen (*Chimaphila maculata*) has leathery, dark evergreen leaves with a clear white midrib. Leaves are found in pairs arranged in whorls. The nodding white flowers have a pinkish tinge and measure ⅔ to 1 inch across. They have a slightly sweet fragrance. It grows 3 to 4 inches tall in dry woodlands in all eastern states and is occasionally found in the mountains of Arizona.

MEDICINAL USE: Both American Indians and European settlers used the leaf as a general tonic and an antiseptic. A poultice was used to treat ringworm, ulcers, and pain. The Shawnee used it to treat consumption. The Pennsylvania Germans believed it would induce sweating to break a chill. Doctors during the mid-19th century used it for digestive problems and as a general tonic to stimulate the appetite and improve energy levels.

Because the plant is green throughout the year, the genus name, *Chimaphila*, is appropriate as it comes from two Greek words meaning "love" and "winter." The common name comes from the Cree word, *pipsisikweu*, which means "breaks into small pieces," and it refers to the use of the plant to break down kidney or bladder stones.

European settlers mixed pipsissewa with mullein and gave it to children to help stop bedwetting. They also used the leaves for flavoring root beer, a practice that continues even today. It is said to have a flavor that is sweet but not minty.

This plant has been known by a multitude of common names, including ratsbane, which is based on the Southern Appalachian belief that it was useful in keeping away rodents. Other names include prince's pine, love-in-winter, noble pine, ground holly, pine tulip, and king's cure.

POKEWEED

Phytolacca americana

TOXIC

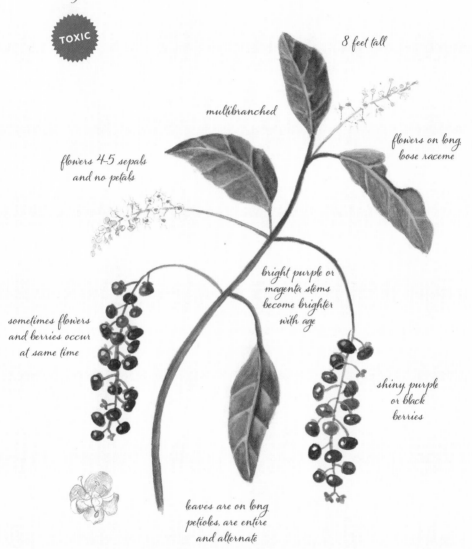

8 feet tall

multibranched

flowers on long, loose raceme

flowers 4-5 sepals and no petals

bright purple or magenta stems become brighter with age

sometimes flowers and berries occur at same time

shiny, purple or black berries

leaves are on long petioles, are entire and alternate

BLOOMS: June–August.

HABITAT AND RANGE: Found commonly in waste places, open fields, and meadows. Native to eastern North America and in Washington State, south to California, and in New Mexico and Arizona.

WILDLIFE PARTNERS: Birds such as gray catbird, northern mockingbird, cardinal, brown thrasher, mourning dove, robin, woodpecker, flicker, and cedar waxwing eat the seeds. It provides food and an important nesting habitat for quail. Small mammals such as raccoons, opossums, red and gray foxes, and the white-footed mouse also eat the berries.

CONSERVATION: Common and abundant.

IN THE GARDEN: It's toxic and weedy and potentially invasive, but it can make a bold statement in the fall. The colors look wonderful with fall-blooming plants such as ironweed and goldenrod. It's easy to grow but can be invasive.

MEDICINAL USE: The Cherokee, Lenape, Iroquois, Pawnee, and Seminole used it medicinally, mostly as a dermatological aid and as a treatment for rheumatism. The mashed root was made into a salve to treat burns and sores. All parts of the plant contain toxins. *Toxic.*

In spite of its toxicity, which increases as the plant matures, the new leaves have been enjoyed and celebrated as a delicious spring green in Southern Appalachia for centuries. To be eaten safely, the leaves must be boiled, and the water discarded at least three times. The resulting dish is called "poke sallet." The new young stalks are also sometimes pickled and eaten.

The toxins of pokeweed can be absorbed through the skin, and you should wear gloves if touching the plant. The old-fashioned antidote for poke poisoning was to drink a lot of vinegar and eat a pound of lard!

The genus name, *Phytolacca*, is from two words, the Greek word *phyton,* meaning "plant" and the Latin word *lacca*, meaning "red dye." Pokeweed ink can be made by mashing the berries (be sure to wear gloves), then straining the seeds out of the juice. The ink doesn't last long, however, even when protected from sunlight.

An old mountain superstition suggested that wearing a string of pokeberries around your neck kept you from catching a contagious disease. Another superstition said that feeding mashed roots and leaves to chickens would make them livelier.

In the presidential campaign of 1844, James K. Polk and his supporters wore pokeberry leaves as a campaign symbol.

SEGO LILY

Calochortus nuttallii

bell-shaped
flowers,
3 inches across

10–20 inches tall

3 sepals, often
purplish-yellow

anthers yellow
or pink

leaves wither
as season
progresses

3 white (lilac, or
pink) petals with
yellow or dark
purple at throat

slender stem
and linear
grasslike leaves

seedpod

BLOOMS: May–June.

HABITAT AND RANGE: The sego lily (*C. nuttalii*) is native to central and western mountain states, usually at elevations between 4,500 to 8,000 feet, on mesa slopes, in brushy, open pine forests, or in high desert areas.

CONSERVATION: Abundant and secure.

WILDLIFE PARTNERS: It is considered an important pollinator plant for native bees, beetles, and butterflies. Pocket gophers, moles, and other rodents eat the seeds.

IN THE GARDEN: Difficult to cultivate. Best grown in very deep, well-drained sandy soils.

RELATED SPECIES: Gunnison's mariposa lily (*Calochortus gunnisonii*) is a stunningly beautiful species found in Colorado and surrounding states. Petals can be pale or deep lavender. Winding mariposa lily (*C. flexuosus*) is also native to Colorado.

Sego lily, in times of abundant rainfall, can experience a "super bloom," meaning that huge colonies form. This happened near Glen Canyon, Utah, in May 2019.

Sego lily is the state flower of Utah. It was credited with saving the Mormons from starvation in the early 19th century when an infestation of crickets ruined all the food crops.

Although the bulbs are, indeed, edible, they are considered a last resort food source as they have little flavor. The bulbs were roasted, baked, or made into porridge. Both the seeds and the flowers are also edible. The seeds were ground into a powder and mixed with other meal for making cakes or biscuits.

The Navajo and Hopi considered it food for children. The raw roots were scooped out and filled with a sweetener, such as sugar or honey, and eaten like candy.

The Hopi also used the flowers in ceremony. In spring, boys collected bouquets of sego lily and larkspur. It was customary for the girls to chase them to try to capture the flowers.

This plant is also called mariposa lily because the flowers look like butterflies. *Mariposa* is Spanish for butterfly.

SNAKEROOT, WHITE

Ageratina altissima (formerly known as Eupatorium rugosum)

TOXIC

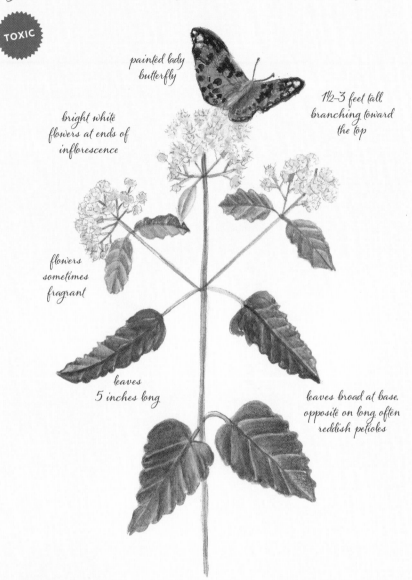

painted lady
butterfly

bright white
flowers at ends of
inflorescence

1½–3 feet tall
branching toward
the top

flowers
sometimes
fragrant

leaves
5 inches long

leaves broad at base,
opposite on long, often
reddish petioles

BLOOMS: Late summer–fall.

HABITAT AND RANGE: Found in open forests, woodland borders, roadsides, and the edges of fields. Native to eastern parts of Canada and the United States, west through the Dakotas, and south to Texas.

WILDLIFE PARTNERS: Flowers attract leaf-cutting bees, wasps, flies, butterflies (monarch, painted lady, sachem, and fiery skipper), and moths. This plant supplies larval food for several caterpillars.

CONSERVATION: Abundant and secure.

IN THE GARDEN: Easy to grow with semi-shade, the white blossoms make a nice bright spot in the fall garden. Just remember that it's toxic to both people and animals.

MEDICINAL USE: As suggested by the common name, snakeroot was at one time used by several American Indian tribes to treat snakebites. The smoke resulting from burning the green leaves was used to revive people who had fainted. *Toxic.*

Cattle that eat this plant pass on the toxins in both their meat and milk. The result of this is "milk sickness," which can be fatal to humans. It's thought that Abraham Lincoln's mother, Nancy Hanks Lincoln, died from milk sickness. At the beginning of the 19th century, when many Western settlers began farming and grazing new lands, thousands of people were reportedly killed by milk sickness. The cause of the illness remained unknown until 1830, when Dr. Anna Pierce Hobbs Bixby, with the help of an unnamed Shawnee woman, traced the poison back to white snakeroot.

The genus name is from the Greek *Ageras,* meaning "not aging" and refers to the flower that keeps its bright white color for a long time. The bright flowers help distinguish them from eupatoriums that bloom at the same time, but with flowers that are more cream-colored.

Eupatorium

Snakeroot

SOLOMON'S SEAL

Polygonatum biflorum

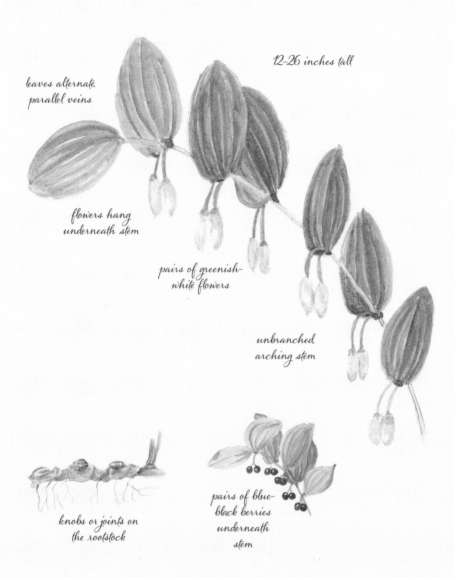

12-26 inches tall

leaves alternate,
parallel veins

flowers hang
underneath stem

pairs of greenish-
white flowers

unbranched
arching stem

knobs or joints on
the rootstock

pairs of blue-
black berries
underneath
stem

BLOOMS: May–June.

HABITAT AND RANGE: Solomon's seal is common in dry or damp woods throughout the range. It grows from Canada south to Florida in the East and as far west as New Mexico and North Dakota.

CONSERVATION: Common and abundant throughout the range.

WILDLIFE PARTNERS: Pollinated by bees and occasionally the ruby-throated hummingbirds. The greater prairie chicken and other woodland bird species eat the berries. Deer and other mammals eat the leaves.

IN THE GARDEN: Prefers a shady site with slightly acidic soils. It benefits from a layer of mulch, particularly in colder regions. False Solomon's seal is also a nice garden plant, preferring loamy soils and partial shade.

RELATED SPECIES: False Solomon's seal (*Maianthemum racemosum*, formerly known as *Smilacina racemosa*) has a terminal spike of flowers and grows 8 to 36 inches tall. Two rows of leaves have parallel veins, are pale green, and are slightly downy on the underneath side. Fall berries are bright red and conspicuous at the end of the stems. Native to much of the country from Alaska south to Arizona and California in the West, and from Nova Scotia south to Georgia and Texas in the East.

MEDICINAL USE: American Indian tribes used the pulverized root of Solomon's seal on wounds to take the color out of a bruise. Tea made from the dried leaves was used as a contraceptive.

The genus name, *Polygonatum,* comes from Greek words that mean "many jointed," and it refers to the number of knobs or joints on the rootstock. Each year the plant sends up a stalk from a new joint, and so the age of the plant can be determined by the number of seals on the root. To some, they resemble a king's seal. King Solomon was said to have been knowledgeable about medicinal plants and put his "seal of approval" on this plant.

American Indians crushed the roots to make flour or sometimes pickled the roots. The black berries of true Solomon's seal are not considered edible, though the red berries of false Solomon's seal can be eaten.

A drawing of the European Solomon's seal was found in a 16th-century herbal and was thought to rid a house of snakes and spiders.

STRAWBERRY, WILD

Fragaria virginiana

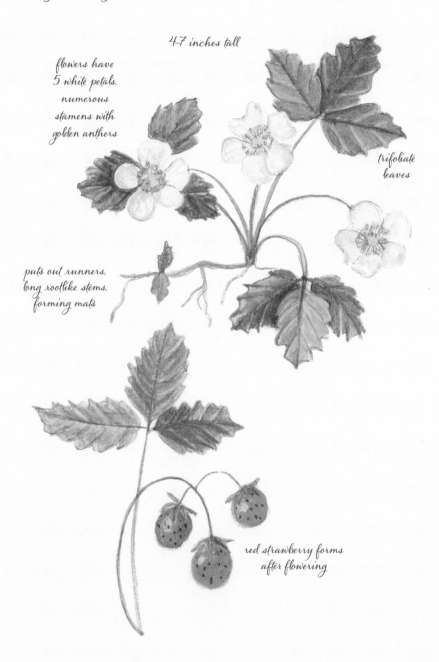

4-7 inches tall

*flowers have
5 white petals,
numerous
stamens with
golden anthers*

*trifoliate
leaves*

*puts out runners,
long rootlike stems,
forming mats*

*red strawberry forms
after flowering*

BLOOMS: Late spring–early summer.

HABITAT AND RANGE: Open fields, meadows, and wood edges. It is native to or naturalized throughout North America.

WILDLIFE PARTNERS: The flowers provide nectar and pollen. They are important plants for native bees, honeybees, flies, small butterflies, skippers, and beetles. The leaves are larval hosts for the grizzled skipper and the gray hairstreak. Birds such as the ring-necked pheasant, brown thrashers, eastern towhees, veeries, and American robins, as well as the ornate box and wood turtles, eat the berries.

CONSERVATION: Abundant.

IN THE GARDEN: Wild strawberry likes full sun and moist conditions. It makes a great ground cover in a sunny area with the added bonus of delicious berries! A cultivar developed from *F. chiloensis*, 'Lipstick', has bright pink flowers. Sometimes called ornamental strawberry, it is used as an evergreen ground cover.

RELATED SPECIES: Hillside or woodland strawberry (*Fragaria vesca*) is native throughout the United States except in Nevada and the Southeast. Beach strawberry (*F. chiloensis*) is native to coastal dunes in Pacific Coast regions, from Alaska south to central California. This latter species and the wild strawberry (*F. virginiana*) are the parents of the commercially cultivated strawberry.

Though it's commonly believed that the name strawberry came from the idea that straw was used as mulch wherever this plant is grown, there is another possible explanation. The Anglo-Saxon word for this plant is *strewberige,* and it was possibly given to the genus because of the great number of runners that are "strewn" all over the ground.

American Indians ate the fruit raw, or they dried it to make pemmican. A tea made from the leaves was thought to stimulate the appetite and clean teeth. It is very high in vitamin C and was used as a general tonic. Strawberry juice mixed with water was called strawberry moon tea.

Don't confuse this plant with the European yellow-blossomed Indian strawberry (*Duchesnea indica*), which produces a very bland, strawberry-looking fruit.

WATER LILY, AMERICAN

Nymphaea odorata

flowers and leaves
float on water surface
(unlike lotus)

leaves round
waxy, deeply cut
10 inches

flower
radially
symmetrical

up to 25 white
(pink) petals with
many yellow
stamens

after pollination
seedpod pulled
underwater until
mature

flowers close by
noon (later in
cloudy weather)

rooted rhizomes,
petioles can grow in
8-feet-deep water

fruit is green, berrylike,
and spongy

BLOOMS: March–October.

HABITAT AND RANGE: The American water lily grows in shallow lakes, ponds, and permanent, slow-moving water. The plant grows in every state except Alaska, Hawaii, North Dakota, and Wyoming. There is some question as to whether or not it is native or naturalized in many of the western states.

CONSERVATION: Abundance varies with location. Though considered an invasive weed in California and Washington, it is considered threatened in Connecticut.

WILDLIFE PARTNERS: Muskrats, porcupines, beaver, moose, and deer eat the leaves. Waterfowl and snapping turtles eat the seeds. It provides a valuable habitat for mink frogs, largemouth bass, and sunfish; it provides a resting platform for damselflies, dragonflies, and frogs. It is pollinated by beetles, which are trapped and drowned in the process.

IN THE GARDEN: Because it is such a beautiful plant, the American water lily has been widely cultivated for water gardens. It needs full sun.

MEDICINAL USE: Many American Indian tribes used the rhizomes to treat coughs, colds, bronchitis, and tuberculosis. The Chippewa pulverized the root and used it to treat mouth sores. The stem, mashed and applied directly to a tooth, was thought to help with toothaches.

All parts are edible: seeds, leaves, flowers, buds, and rhizomes. The unopened buds were considered especially tasty both by American Indians and European settlers. The nutritious seeds were fried or roasted and eaten like popcorn, or they were ground into a powder that was used for thickening soups. American water lily is still considered a favorite wild plant treat.

This plant is also known as fragrant water lily or beaver root. The genus, *Nymphaea*, refers to the habitat, where you would, supposedly, also find water nymphs.

YARROW

Achillea millefolium

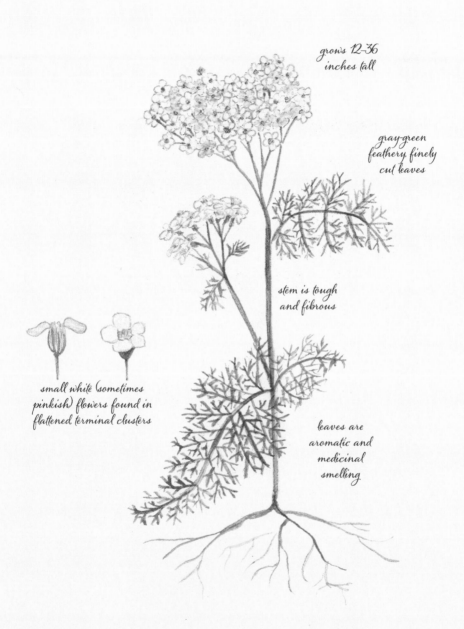

grows 12–36 inches tall

gray-green feathery, finely cut leaves

stem is tough and fibrous

small white (sometimes pinkish) flowers found in flattened terminal clusters

leaves are aromatic and medicinal smelling

BLOOMS: March–October.

HABITAT AND RANGE: A common and abundant plant throughout temperate North America, Europe, and Asia, it is most often found in fields and along roadsides.

CONSERVATION: Abundant.

WILDLIFE PARTNERS: Yarrow attracts many different kinds of butterflies. It is a host plant for many kinds of caterpillars, particularly those of moths, including the striped garden, blackberry looper, cynical Quaker, and voluble dart.

IN THE GARDEN: Yarrow has been extensively bred and cultivated so that now there are vibrantly colored flowers available in shades of gold, pink, red, lavender, light yellow, and dark pink. The cultivated yarrows sometimes have weak and spindly stems, causing them to flop over. Full sun helps produce strong stems. All yarrows tend to spread rapidly and may be a little invasive in the garden.

MEDICINAL USES: Chemicals in the plant are effective in clotting blood. The Crow made salves and poultices by crushing boiled leaves and mixing them with goose grease to treat wounds, burns, boils, earaches, and skin sores. The Cherokee and Cheyenne made a tea from leaves to drink to help induce sweat to break a fever. The Blackfoot used it externally for sore muscles. Western settlers chewed on leaves to settle an upset stomach or help regulate menstrual flow.

The medicinal uses of this plant led to many different common names, including soldier's woundwort, nosebleed, bloodwort, and staunchweed. The species name, *millefolium*, is literally translated to mean "a thousand leaves" and is descriptive of the feathery foliage.

The genus name, *Achillea,* is from the Greek hero Achilles. According to legend, he used this plant to treat wounded soldiers during the Trojan War. Perhaps due to this legend, yarrow became a symbol of war, and the ancients referred to it as *herba militaris.*

The medicinal usage of yarrow predates the Trojan War by tens of thousands of years. Studies done on the teeth of Neanderthal people indicate that they had ingested both chamomile and yarrow. Because yarrow is very bitter, it's presumed that they took it as a medicine, not for food. Yarrow was also found in a cave in Spain dating back fifty thousand years.

According to a Chinese proverb, yarrow brightens the eye and promotes intelligence. Dried yarrow stalks have been used traditionally with the Chinese book, *I Ching,* for divination since the 12th century: fifty yarrow stalks are thrown to get a reading. It is still possible to purchase yarrow for this purpose. In both classic and traditional Chinese medicine, yarrow is used to stimulate chi.

This plant was used to flavor beer and ale in England for centuries.

Yarrow is toxic to dogs, cats, and horses, though instances of actual poisoning are quite rare, as the animal would have to ingest a large amount of these bitter leaves to be harmful.

YUCCA

Yucca filamentosa

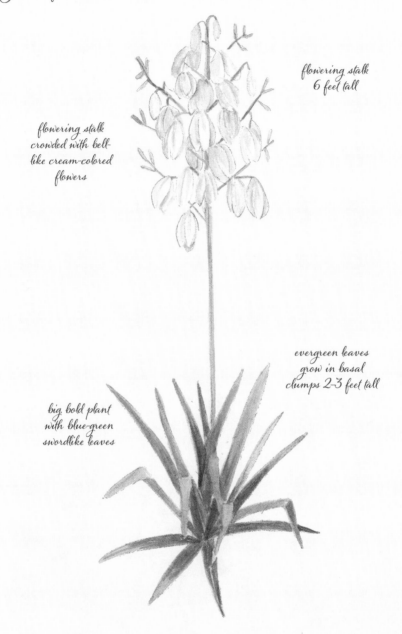

flowering stalk
6 feet tall

flowering stalk
crowded with bell-
like cream-colored
flowers

evergreen leaves
grow in basal
clumps 2-3 feet tall

big, bold plant
with blue-green
swordlike leaves

BLOOMS: April–August.

HABITAT AND RANGE: *Y. filamentosa* is the most common of all the yuccas, growing throughout the East from New York, south to Florida and Alabama, and naturalized throughout. It prefers dry, open woods, fields, or prairies.

CONSERVATION: Many species of yuccas are pro-tected in the states in which they grow. Though the Joshua tree (*Yucca brevifolia*) was considered endan-gered, more relaxed requirements deemed this unnec-essary. This species provides food and shelter for many bird and insect species.

WILDLIFE PARTNERS: The yucca plant and the yucca moth are essential to one another. The plant provides food and habitat throughout the life cycle of the moth, and the moth cross-pollinates the plant. In addition, jackrabbits, desert cottontails, and wood rats eat the sharp, waxy leaves of some species. Gophers eat the roots. Song-birds and game birds eat the blossoms, fruits, and seeds. Antelope, mule deer, and elk eat the flowering stalk. The soaptree yucca provides a nesting site for the cactus wren and protection for the Swainson's hawk and Aplomado falcon.

IN THE GARDEN: Many different kinds of yuccas make superb focal points in the garden. They are relatively easy to grow, needing dry conditions and well-drained soils. Because of their dangerously sharp leaves, plant them in out of the way places in the garden.

RELATED SPECIES: One of the most beau-tiful western yuccas is banana yucca (*Yucca bac-cata*). It looks similar to *Yucca filamentosa*, but the flowering stalk of *Y. baccata* is about the same length as the leaves. It grows in many places of the Southwest, north to Utah.

MEDICINAL USE: The Cherokee used the roots for treating sores and rashes. The

Catawba used the root to treat skin ailments. Other tribes used yucca for treating migraines, high blood pressure, stomach ailments, and diabetes.

There are between 40 to 48 species of yuccas, most of which are native to the Americas. Indigenous peoples have always used yucca for a variety of purposes. The inner part of the leaves provided filaments used for weaving baskets or making sandals. Several species were used as a strong soap for washing and as a shampoo, which was good for getting rid of dandruff.

The buds and blossom, picked at the right time, were thought to be flavorful. The fruit was cooked and the seeds removed before eating.

The Navajo used the leaves in ceremony, and the Pueblo had a Yucca Dance as part of their tribal tradition.

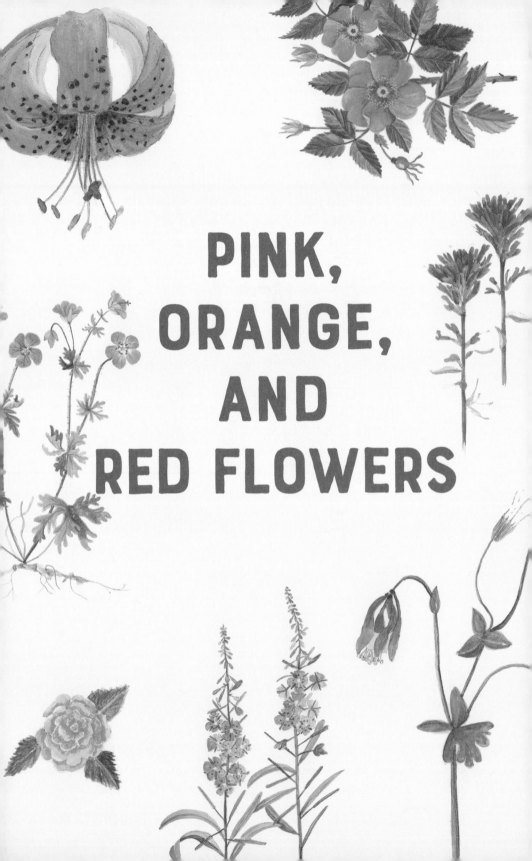

PINK,
ORANGE,
AND
RED FLOWERS

Bee balm *Monarda didyma*

Bitterroot *Lewisia rediviva*

Columbine, eastern *Aquilegia canadensis*

Coneflower, purple *Echinacea purpurea*

Elephant head *Pedicularis groenlandica*

Fire pink *Silene virginica*

Fireweed *Chamaenerion angustifolium*

Geranium, wild *Geranium maculatum*

Indian paintbrush *Castelleja coccinea*

Joe-pye weed, spotted *Eutrochium maculatum*

Lady's slipper, pink *Cypripedium acaule*

Lily, wood *Lilium philadelphicum*

Milkweed, common *Asclepias syriaca*

Monkey flower, Lewis's *Erythranthe lewisii*

Penstemon, Parry's beardtongue *Penstemon parryi*

Phlox *Phlox paniculata*

Pitcher plant, purple *Serracenia purpurea*

Prairie smoke *Geum triflorum*

Rose, wild *Rosa* species

Shooting star *Dodecatheon pulchellum*

Skyrocket *Ipomopsis aggregata*

Trillium *Trillium* species

Vetch, American *Vicia americana*

BEE BALM
Monarda didyma

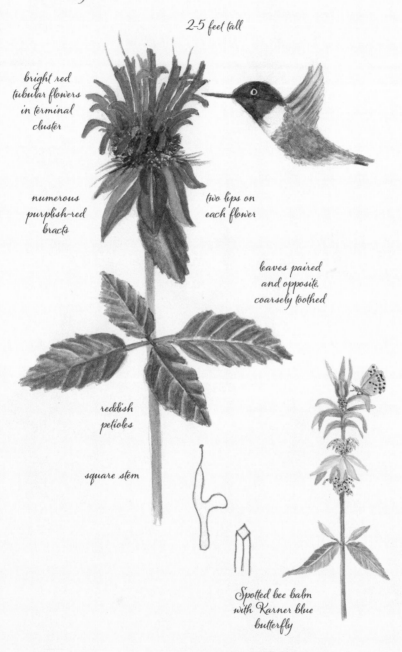

2-5 feet tall

bright red
tubular flowers
in terminal
cluster

numerous
purplish-red
bracts

two lips on
each flower

leaves paired
and opposite,
coarsely toothed

reddish
petioles

square stem

Spotted bee balm
with Karner blue
butterfly

BLOOMS: Early summer.

HABITAT AND RANGE: Stream banks or moist woods, eastern North America from New York to Georgia and Tennessee to Michigan.

CONSERVATION: Spotted bee balm (*Monarda punctata*) is considered endangered in Ohio and Pennsylvania. It is especially important for the endangered Karner blue butterfly. It has petal-like bracts that occur in pink, white, purple, or yellow. The true flowers are small, green, and covered with purple spots, thus the common name. It is an annual that grows throughout the eastern United States and prefers sandy soils.

WILDLIFE PARTNERS: Considered an important plant for pollinators, bee balm is pollinated by ruby-throated humming-birds and butterflies such as swallowtails and fritillaries. Hawk moths, bumble-bees, and sand wasps (which eat stink bugs!) are also attracted to this plant.

IN THE GARDEN: Perennial, cultivated monardas in the garden come in different colors, including purple and pink. They prefer full sun, soils that are well-drained and moist, and good air circulation. Deadhead for repeat blooming and cut back to 2 inches above the soil after the first frost in fall.

MEDICINAL USES: The Cherokee made tea from bee balm for its soothing qualities. It was drunk to improve digestion and to reduce nausea and vomiting. A poultice from mashed leaves was applied to the skin to ease pain and itching from bee stings and insect bites.

The genus, *Monarda,* is named for a Spanish physician, Nicolás Monardes, who wrote several books in the 16th century introducing North American plants to Europe. The species name, *didyma,* is from the Greek word meaning "paired" and refers to the two lips found on the flower. It was also known as wild bergamot because the smell is similar to the cultivated orange bergamot.

This plant is also known as Oswego tea, named for American Indians of the Oswego area of New York, who made a tincture from the leaves and blossoms. This tea is still a popular floral treat.

Bee balm is quite tasty. Use both the leaves and blossoms raw as a garnish or in salads, or to decorate cakes and cupcakes. In addition, dried leaves and flowers can be made into potpourris.

BITTERROOT

Lewisia rediviva

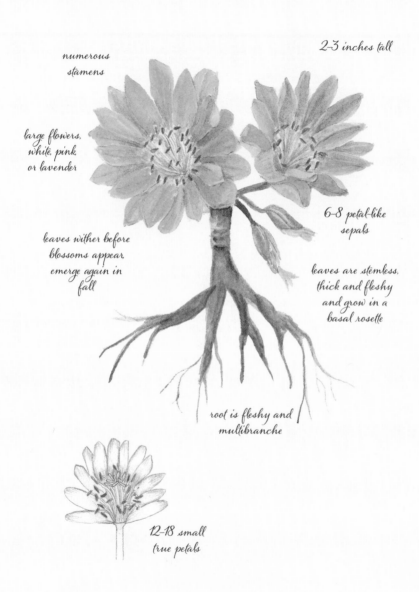

numerous
stamens

2-3 inches tall

large flowers,
white, pink
or lavender

leaves wither before
blossoms appear,
emerge again in
fall

6-8 petal-like
sepals

leaves are stemless,
thick and fleshy
and grow in a
basal rosette

root is fleshy and
multibranche

12-18 small
true petals

BLOOMS: April–July.

HABITAT AND RANGE: Grows in dry, gravelly, sandy open areas with little other vegetation from Washington to California, east to Montana, Colorado, and Arizona. From elevations of 2,500 to 10,000 feet.

CONSERVATION: Bitterroot is abundant. Sacajawea's bitterroot (*Lewisia sacajaweana*) is found only in central Idaho and is considered "imperiled." This is the first plant named after the Lemhi Shoshone woman, Sacagawea, who served as a guide for the Lewis and Clark Expedition.

WILDLIFE PARTNERS: Leaves and seeds are eaten by rodents and by the gray-crowned rosy finch. Bitterroot flowers attract butterflies and bees.

IN THE GARDEN: Many different cultivars have been developed from the related Siskiyou bitterroot (*L. cotyledon*). The flowers are orange, gold, peach, pink, white, or striped, and the plant is considered easy to grow if given perfect drainage and is not overwatered.

MEDICINAL USES: The Flathead made tea from the roots for heart trouble and pleurisy. The Nez Perce used the tea to help milk flow for nursing mothers.

The genus is named for Meriwether Lewis, who collected it in Montana (in what is now the Bitterroot Valley) in 1806. The Bitterroot Mountains and River were also named for this plant. There are 19 species of *Lewisias,* and all are native to the western United States. The species name, *rediva,* means "to revive," so named because of its amazing ability to rejuvenate. Plants dried for over a year will revive after a good soaking.

Bitterroot was the most important root crop of the Flathead and the Kootenai in western Montana. Each spring, tribes gathered to dig the root in areas where it was abundant. The season began with the First Roots Ceremony. Women were primarily the ones to dig the roots, only accompanied by men who served to protect them from wild animals and neighboring tribes.

A special digging stick was made for extracting the root. It was usually made from willow, with one end sharpened and hardened by fire and the other end fixed with an antler handle.

The roots were such an important food for these tribes that they were used as currency. It took a woman three to four days to fill a 50-pound sack of roots, and this was said to be equal in value to a horse.

The Lemhi Shoshone believed that the small red core in the upper taproot had special powers and could stop a bear attack.

This is the state flower of Montana.

COLUMBINE, EASTERN

Aquilegia canadensis

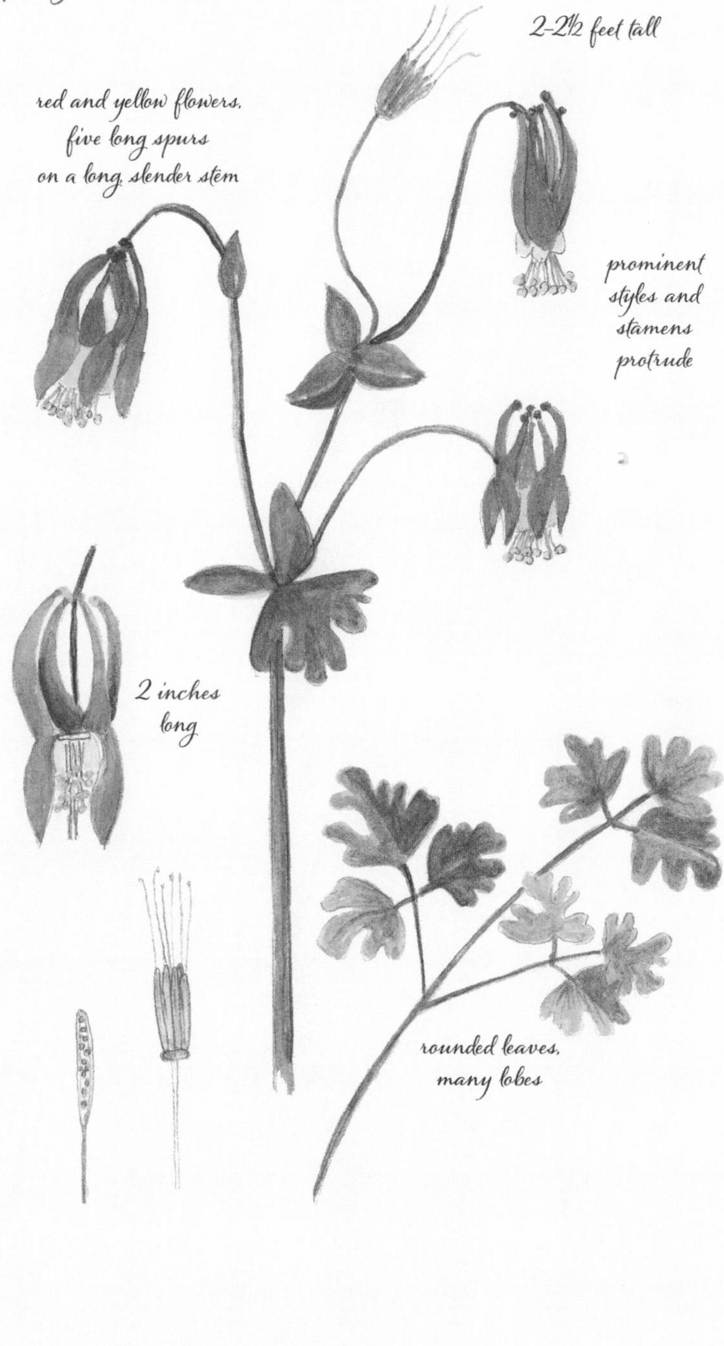

2-2½ feet tall

red and yellow flowers,
five long spurs
on a long, slender stem

prominent
styles and
stamens
protrude

2 inches
long

rounded leaves,
many lobes

BLOOMS: March–April.

HABITAT AND RANGE: Grows on rocky slopes, dry and open areas, and open woods. It is found in eastern North America from Canada south to Florida and west to Arkansas and Texas.

CONSERVATION: Common and abundant.

WILDLIFE PARTNERS: Different species of columbines attract hummingbirds throughout their range, and they are a food source for many species of butterfly and moth caterpillars. Yellow columbine (*Aquilegia chrysantha*), which is native to the American Southwest, is generally pollinated by hawk moths.

IN THE GARDEN: The western columbine (*A. formosa*) looks very much like its eastern cousin, and their growing requirements are similar. They both need well-drained, rich soils and sun to light shade. They are both short-lived perennials, but both reseed readily. Most cultivars have been developed from the European columbine (*A. vulgaris*), and the flowers come in blue, pink, white, orange, red, and yellow. They need full sun and well-drained soil, and they benefit from good air circulation as they are subject to mildew.

MEDICINAL USES: In 1373, European columbines were used with seven other herbs as a cure for the "pestilence." It was used for treating abdominal pain, measles, smallpox, and swelling. Taken with saffron, it was used to cure jaundice.

Paleontologists have determined that during the Pleistocene epoch, when the Bering land bridge connected America and Asia (about 10,000–40,000 years ago), a species of columbine crossed into North America. As it spread from Alaska throughout the continent, it adapted to local pollinators and developed into new species. Some flowers were more successful with their spurs pointing upward, while others found better pollination with the spurs downward. Even today, you can easily see the correlation between the pollinators and the flowers, based on the length of the spurs.

Long-spurred columbines grow primarily in the Southwest, where hawk moths are common. Red and yellow columbines, which have medium-length spurs, only grow in the East where the ruby-throated hummingbird is the predominant pollinator. Short-spurred columbines are pollinated by bumblebees, which have very short tongues.

Many of the names for columbines, both scientific and common, refer to the flowers' spurs. The common term *meeting house* refers to the upward-pointing spurs that look like heads sitting in a circle; the term *columbine* comes from the Latin word for dove (*columba*), because the spurs look like doves' heads; the genus name, *Aquilegia,* comes from the Latin word for eagle because the downward-pointing spurs look like an eagle's claw.

Columbine blossoms were used as flavoring, and a 1494 banquet menu listed "gely coloured with columbine floures." John Gerard, an ancient herbalist, called it *Herba Leonis,* or the "herbe wherein the lion doth delight." This referred to the old superstition that lions ate columbines in spring to revive their strength.

An old European superstition said that to carry columbine in your pocket would bring you good luck.

CONEFLOWER, PURPLE

Echinacea purpurea

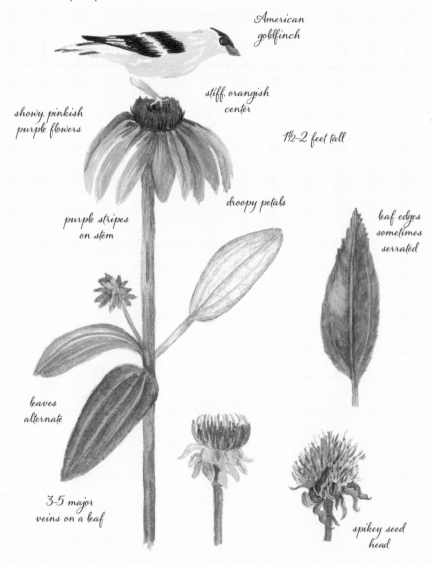

American goldfinch

stiff, orangish center

showy, pinkish purple flowers

1½–2 feet tall

droopy petals

purple stripes on stem

leaf edges sometimes serrated

leaves alternate

3–5 major veins on a leaf

spikey seed head

BLOOMS: June–September.

HABITAT AND RANGE: Found in sunny, open spaces in the Northeast, Southeast, Midwest, and north into Canada.

CONSERVATION: Though a common garden plant, purple coneflower is not often found growing in the wild. The endangered smooth coneflower (*E. laevigata*) is found only in two Georgia counties and a few other places in the Southeast.

WILDLIFE PARTNERS: Attracts butterflies such as monarchs, painted ladies, fritillaries, and swallowtails. Birds such as goldfinches, cardinals, and blue jays eat its seeds. Deer may eat the young plants but usually leave mature plants alone.

IN THE GARDEN: Perennial purple coneflower is an easy to grow, highly successful garden plant. Cultivars have been developed that display many flower colors, including white, cream, yellow, orange, red, violet, and purple. It needs full sun, well-drained soils, and good air circulation. It tolerates drought, poor soils, heat, and humidity. It spreads by runners and will reseed readily.

RELATED SPECIES: Narrow-leaved purple coneflower (*Echinacea angustifolia*) looks very much like purple coneflower (*E. purpurea*), though the leaves are smaller. Most of the developed cultivars come from *E. angustifolia,* and it is considered more effective medicinally.

MEDICINAL USES: Echinacea has been important in folk medicine wherever it grows. The Lenape used the root as a treatment for gonorrhea. The Plains Indians used it as a general tonic and specifically as a painkiller for toothaches, coughs, colds, sore throats, and snake bites. The Choctaw used it as cough medicine and as a gastrointestinal aid. Echinacea was first discovered growing in the southeastern United States, and in the late 17th century, it was sent to England as a garden plant. Once established in Europe, it was used during the 18th and 19th centuries to treat scarlet fever, syphilis, malaria, and diphtheria.

Today this is a very popular alternative medicine because of its supposed effectiveness in boosting the immune system. Specific dosage is important when using this supplement, and there is actually no clear proof of its effectiveness.

ELEPHANT HEAD

Pedicularis groenlandica

MILDLY TOXIC

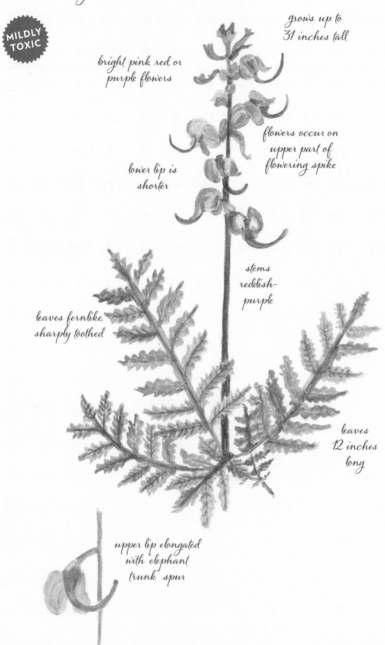

grows up to 31 inches tall

bright pink red or purple flowers

flowers occur on upper part of flowering spike

lower lip is shorter

stems reddish-purple

leaves fernlike, sharply toothed

leaves 12 inches long

upper lip elongated with "elephant trunk" spur

BLOOMS: June–August.

HABITAT AND RANGE: Found in wet areas such as bogs, marshes, and wet meadows, particularly at higher elevations. It grows throughout the western half of the United States.

CONSERVATION: Locally abundant throughout the range.

WILDLIFE PARTNERS: Pollinated by at least five different kinds of bumblebees, both queens and workers, and, later in the season, by hummingbirds.

IN THE GARDEN: This plant is a perennial; it needs high moisture and full sun. Sow seeds in the fall.

MEDICINAL USES: The Cheyenne used the powdered or dried leaf to treat coughs. Indian warrior or warrior's plume (*Pedicularis densiflora*) is native to California and Oregon, and it is used even today as a sedative or tranquilizer when treating anxiety, tension, and insomnia. *Mildly toxic.*

The unusual shape of the flower, which really does look like the head of an elephant, has given rise to other common names such as bull elephant head or elephanthead lousewort.

Carolus Linnaeus, the "father of modern botany," placed this plant in the lousewort family. It was once thought that members of this family would either spread lice to humans and animals or that the plant was used to treat an infestation of lice, neither of which has proven to be true. The species name, *groenlandica,* means "of Greenland" where it also grows.

Many species of *Pedicularis,* including this one, are hemiparasitic on other plants growing close by. The roots penetrate host plants and absorb nutrients from them.

Pollination for *P. groenlandica* varies during the flowering season. The short, early-season flowers are pollinated by short-tongued bumblebees such as *Bombus edwardsii.* As the season progresses and the flowers continue to grow and lengthen, they are more often pollinated by hummingbirds such as Anna's and rufous.

FIRE PINK

Silene virginica

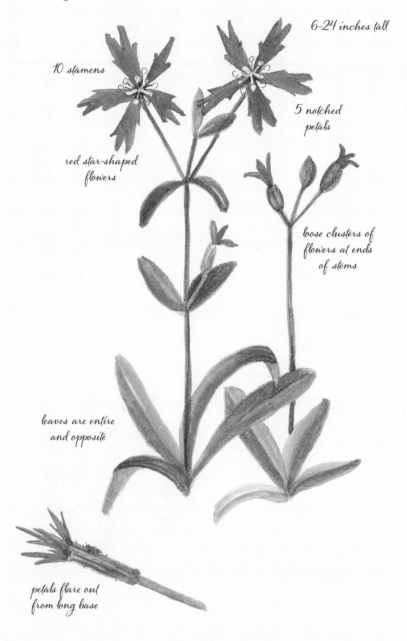

6–24 inches tall

10 stamens

5 notched petals

red star-shaped flowers

loose clusters of flowers at ends of stems

leaves are entire and opposite

petals flare out from long base

BLOOMS: April–mid-summer.

HABITAT AND RANGE: This species prefers open woods or rocky slopes. Though somewhat uncommon, it is widespread throughout the East, and west of the Mississippi River to Oklahoma and Kansas.

CONSERVATION: Fire pink (*S. virginica*) is considered threatened in Michigan, and it is endangered in Wisconsin and Florida. Several western species of *Silene* are rare and threatened or endangered.

WILDLIFE PARTNERS: The main pollinator plant for fire pink is the ruby-throated hummingbird.

IN THE GARDEN: Fire pink makes a good garden plant. It grows fairly easily from seed or root cutting. It likes well-drained soils, sunshine, and slightly acidic soils. The plants last only two to three years.

RELATED SPECIES: Several other *Silene* species are native to the western states. These include Douglas's catchfly (*Silene douglasii*), which has a blossom that resembles a bladder, moss campion (*S. acaulis*), Red Mountain catchfly (*S. campanulata*), and cardinal catchfly (*S. laciniata*).

MEDICINAL USES: In Elizabethan England, *Silene* was called gillofloures. It was mixed with sugar and wine to make a concoction good for the heart. American Indian tribes also thought that *Silene* was good for the heart and for soothing anxiety. They used available species as an antibacterial agent, as a remedy to help soothe children with colic, as a medicine to help sick babies, as a treatment for ant bites, and as a cure for the bite of a coyote or prairie dog.

The genus name, *Silene*, may come from the Greek word *sialon*, meaning saliva. The other possibility is that the genus was named for Silenus, the foster father of Bacchus in Greek mythology. Silenus was often pictured with beer all over his face, which perhaps reminded some of the sticky substance all over the stem of the plant.

The common name, "catchfly," is often associated with all members of this family. The name was given because of the sticky sap that is found on the calyx or stem of so many species. This sap helps prevent insects from crawling up the stem to steal the nectar.

FIREWEED

Chamaenerion angustifolium (formerly known as Epilobium angustifolium)

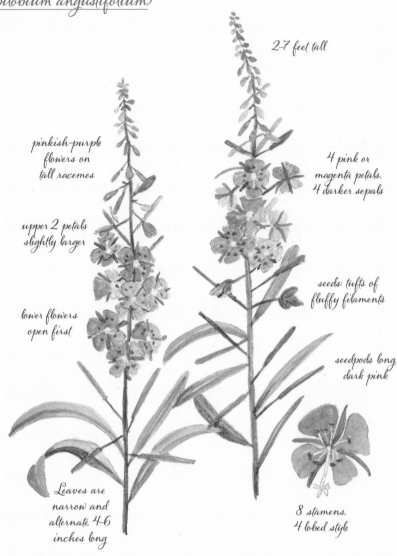

2-7 feet tall

pinkish-purple flowers on tall racemes

4 pink or magenta petals, 4 darker sepals

upper 2 petals slightly larger

lower flowers open first

seeds: tufts of fluffy filaments

seedpods long, dark pink

Leaves are narrow and alternate, 4-6 inches long

8 stamens, 4 lobed style

BLOOMS: June–September.

HABITAT AND RANGE: Native throughout the northern hemisphere in disturbed areas. Fireweed is particularly apparent after forest fires (thus, the common name).

CONSERVATION: Abundant throughout the range.

WILDLIFE PARTNERS: This plant is an import-ant forage plant and a preferred food for deer, moose, caribou, muskrat, and hares. It is the larval host for the white-lined sphinx moth and is pollinated by flies and eight different species of bees.

IN THE GARDEN: This attractive perennial can become an aggressive weed in well-tended gardens. It spreads by underground runners and by seeds dispersed by wind. It grows less aggressively in light shade. This can be started from seed sown directly into the garden in the fall.

MEDICINAL USE: The Abenaki used the dried leaves to make tea for fevers and coughs. The Chippewa put it on bruises and used the macerated dried roots, mixed with bear grease, on bites and sores. The Cree put crushed leaves on burns and rashes. Currently, there is some interest in using fireweed to treat an enlarged prostate.

The common name, fireweed, comes from the fact that this is one of the first and certainly the showiest plants to appear after forest fires. In Britain, fireweed is called willow herb because its leaves look much like willow leaves. Other common names include blooming sally and rosebay.

The Blackfoot removed the inner pith of the stem and rubbed it on their hands and face as protection from the cold or rubbed the flowers on rawhide as waterproofing. The tough fibers were used as thread for making fishing nets.

Inuit ate the raw root and new shoots, while their children ate the sweet pith.. The young leaves, boiled and mixed with other greens and a bit of meat, were also considered tasty.

The seeds have long silky threads at one end and look somewhat like those of milkweed. These were used for stuffing in blankets or pillows and were combined with dog hair to make thread for weaving.

Bees that sip on fireweed blossoms make a light, delicate honey that is considered the "champagne" of honey.

GERANIUM, WILD

Geranium maculatum

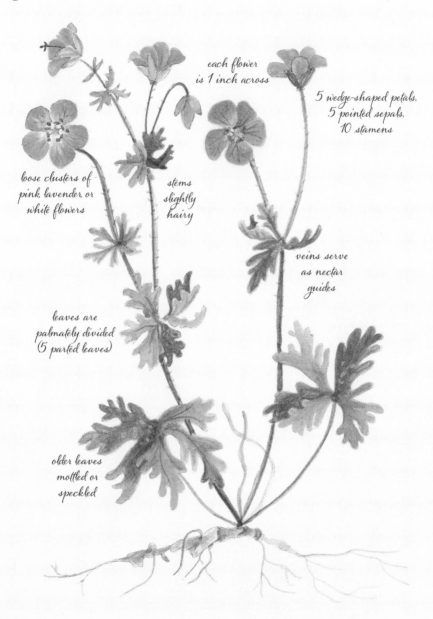

each flower
is 1 inch across

5 wedge-shaped petals,
5 pointed sepals,
10 stamens

loose clusters of
pink lavender or
white flowers

stems
slightly
hairy

veins serve
as nectar
guides

leaves are
palmately divided
(5 parted leaves)

older leaves
mottled or
speckled

BLOOMS: April–June.

HABITAT AND RANGE: Found in open, moist woodlands and shady roadsides throughout the eastern United States, south to Georgia, west to Louisiana, and north to the Dakotas.

CONSERVATION: Haleakala National Park in Hawaii has been called the geranium capital of the world because it is home to four rare and endangered wild geraniums that grow nowhere else. The many-flowered geranium (*G. multiflorum*) is pollinated by the Hawaiian yellow-faced bee; a rare geranium known as *G. hanaense* is found in high-altitude bogs and was discovered in 1988; the silver geranium (*G. cuneatum*) is the most common of these rare species at the park; and the very rare Hawaiian red-flowered geranium (*G. arboreum*) is the only bird-pollinated geranium in the world and is pollinated by the Hawaiian honeycreeper, the i'iwi.

Hawaiian honey-
creeper

WILDLIFE PARTNERS: Geraniums are pollinated by mason bees, flies, dance flies, butterflies, skippers, and beetles.

IN THE GARDEN: Many cultivars have been developed from the wild geraniums, and all make lovely additions to the wildflower garden. 'Rozanne' creates large mounds of bright blue flowers, and the Perennial Plant Association named it the Perennial Plant of the Year in 2008. Provide part sun and well-drained soil for best blooms.

RELATED SPECIES: The geranium known as herb Robert (*Geranium robertianum*) is a European species that has naturalized in many parts of the United States. In Washington, it is considered a noxious weed. When the leaves are crushed, it smells something like burning rubber, resulting in the common name, stinking Bob. No one is quite sure who this species was named after, but some suggest it was Robert Goodfellow, otherwise known as Robin Hood.

MEDICINAL USES: The wild geranium (*G. maculatum*) was included in the *United States Pharmacopeia* from 1820–1916. Tea from dried geranium leaves was used by European settlers for sore throats and mouth ulcers, diarrhea, and dysentery. High levels of tannin make geraniums good for helping blood coagulate, and it was used on wounds or to prevent hemorrhages. American Indians used it as a tonic and astringent and to treat diarrhea. Tea from geranium roots is still used today in combination with other herbs to help increase fertility.

An ancient recipe for hair tonic was made from geraniums, lady's-smocks, and twayblades. It was given to bald men to help hair grow. Crushed geranium leaves were smeared on the skin to help repel mosquitoes.

The seedpods are elongated and slightly curved, and many people have thought that they look like a crane's bill. Both the genus and common names refer to this feature. The Greek word *geranus* means "crane." Other common names include storksbill or crow's foot, which refer to the shape of the leaf.

The word *shameface* comes from a legend that said, at one time, all geraniums were white. Then one day, the prophet Muhammad washed his shirt in a stream and placed it on the geraniums to dry. They blushed pink at this distinction and have been that color ever since.

Great confusion about the difference between the popular bedding plant called geranium and the true wild geranium still lingers. The problem began in the 1600s when plant explorers introduced a new plant from South Africa that looked something like the native geraniums growing in England. They called the new plants geraniums, and the name and the confusion persist, even though the South African plants are actually in the genus *Pelargonium*.

According to the language of flowers, geraniums symbolize constancy in love. A European superstition suggests that carrying a sprig of geranium was thought to bring good luck.

INDIAN PAINTBRUSH

Castilleja coccinea

TOXIC

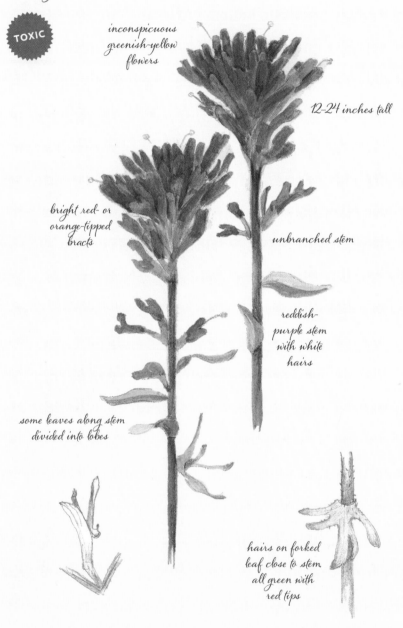

inconspicuous greenish-yellow flowers

12–24 inches tall

bright red- or orange-tipped bracts

unbranched stem

reddish-purple stem with white hairs

some leaves along stem divided into lobes

hairs on forked leaf close to stem, all green with red tips

BLOOMS: April–August.

HABITAT AND RANGE: Indian paintbrush (*C. coccinea*) is found in moist meadows, prairies, and roadsides. It is only one of three species of paintbrush found east of the Mississippi River, and it grows from Maine, west to Minnesota, and south to Louisiana and Georgia. It is also found west of the Mississippi River in Texas, Oklahoma, and Kansas.

CONSERVATION: Christ's Indian paintbrush (*C. christii*) is considered threatened and has been under consideration for the US Endangered Species list.

WILDLIFE PARTNERS: The *Castilleja* genus is an important pollinator plant for many native species, particularly bees. It also attracts broadtail hummingbirds and butterflies such as the Fulvia checkerspot.

IN THE GARDEN: Indian paintbrush is a biennial, producing a rosette of leaves the first year and flowering the second year. Like many species of *Castilleja,* it is difficult to establish in cultivation because it is hemiparasitic, depending on perennial grasses for nutrients. It prefers moist, sandy, well-drained soils. Grow from seeds; do not try to transplant.

RELATED SPECIES: Splitleaf Indian paintbrush (*Castilleja rhexifolia*) features bright pink bracts and small green flowers. Plants show great variation and may or may not be covered with white hairs. Lower leaves are also tinged with pink. It ranges from New Mexico north to Montana, west to Washington, and south to Oregon.

MEDICINAL USES: American Indians used a tea made from the flowers to treat rheumatism. *Toxic.*

The name comes from a legend about a young American Indian boy who prayed to the Great Creator to give him brilliant paint to capture the colors of the sunset. His prayers were answered with paintbrushes growing in the ground. After he used them, he put them back, and they turned into this flower. Other common names are prairie fire and scarlet painted cup.

There are 200 species of Indian paintbrushes. Almost all are native to western parts of North America, although some are in the Andes and some are in Asia. The Wyoming Indian paintbrush (*C. linariifolia*) was named the Wyoming state flower in 1917.

Indian paintbrush absorbs selenium (a potentially toxic metal) from the soil. The Ojibwe used the plant to make a hair wash, thought to be effective because of the high amount of selenium present in the oils of the plant. Indian paintbrush was used as a poison by the Cherokee and as a love charm by the Menominee. It was occasionally eaten by several different tribes as a condiment, flavoring other foods.

"To see a World in a Grain of Sand
And a Heaven in a Wild Flower"
—William Blake,
"Auguries of Innocence"

JOE-PYE WEED, SPOTTED

Eutrochium maculatum (formerly known as Eupatorium maculatum)

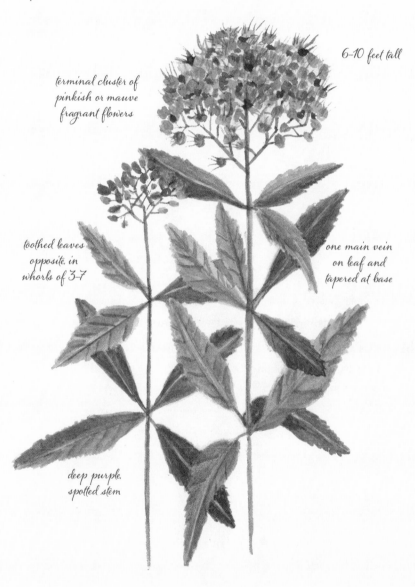

6–10 feet tall

terminal cluster of pinkish or mauve fragrant flowers

toothed leaves opposite, in whorls of 3–7

one main vein on leaf and tapered at base

deep purple, spotted stem

BLOOMS: Late July–September.

HABITAT AND RANGE: Damp meadows, roadsides, and fields throughout eastern states and in many western states.

CONSERVATION: Common and abundant throughout its range.

WILDLIFE PARTNERS: An important source of nectar for bumblebees, honey and digger bees, wasps, flies, moths, and many types of butterflies. It is a favorite plant of the endangered rusty patched bumblebee. Many birds, including the goldfinch, Carolina wren, dark-eyed junco, and the tufted titmouse, eat the seeds.

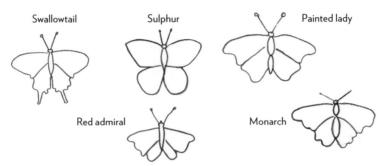

Swallowtail Sulphur Painted lady

Red admiral Monarch

IN THE GARDEN: Needs full sun and prefers soils rich in humus. This plant is easy to grow, but it gets very tall, so place it at the back of a border or bed.

MEDICINAL USES: The Cherokee made a tea from fresh or dried leaves to treat kidney disorders and rheumatism. The Chippewa included it in a bath to soothe nerves, especially for children. The Iroquois used a tea to treat fevers.

Joe Pye was thought to be a *sachem* (chief) in the Mohican tribe who used this plant to treat typhoid fever. Some sources say that he lived from 1740–1785 in Massachusetts, though other sources are more vague as to the exact identity, life span, and relative importance of this man. Many people think there was never a man named Joe Pye and that, instead, the plant name came from an Algonquian word for typhoid, *jopi.* Because this plant was considered a cure for typhoid fever, it was known as jopi weed, which eventually became joe-pye weed. In the Southern Appalachian Mountains, it was also known as queen of the meadow.

American Indians thought this plant would bring good luck in love.

LADY'S SLIPPER, PINK

Cypripedium acaule

6-15 inches tall

looks like a shoe with 2 greenish brown petals

3 sepals

single pink or magenta (rarely white) flower

leafless flower stalk

2 basal leaves with parallel veins

BLOOMS: April–July.

HABITAT AND RANGE: Lady's slipper prefers dry woods or forest hillsides, and it is often found in mossy areas. The range extends from Canada south to Georgia, west to Alabama, and north to Minnesota.

CONSERVATION: Lady's slipper is a subfamily of orchids. Though many native orchids are threatened or endangered (see Orchids, page 104), pink lady's slipper (*C. acaule*) is secure and relatively common.

WILDLIFE PARTNERS: Pollinated primarily by native bees, it has a symbiotic relationship with fungus in the soil. When plants are young, they obtain nutrients from the fungus. As they mature and can produce their own nutrients, they supply food for the fungus.

IN THE GARDEN: A spectacularly beautiful wildflower, lady's slippers are exceedingly difficult to cultivate because of their symbiotic relationship with fungus. This is true for almost all North American species.

RELATED SPECIES: In the West, mountain lady's slipper (*Cypripedium montanum*) is white with a blush of pink and California lady's slipper (*C. californicum*) is white and dull purple. Both are found in areas with rich soils and light shade. They bloom late spring into summer, from May through July. Yellow lady's slipper, *Cypripedium parviflorum*, is found in boggy areas throughout most of North America.

MEDICINAL USES: Algonquian-speaking tribes used lady's slipper to treat menstrual problems, venereal disease, and stomach aches. The Cherokee used it to treat spasms and "fits." The Rappahannock used the dried root, mixed with whiskey, as a general tonic. Many tribes used it for children with kidney problems and for insomnia, tension, hysteria, and anxiety. The Meskwaki used it as a love medicine. Lady's slippers, as well as other North American native orchids, were used in Europe as a sedative and as a substitute for valerian.

The genus name *Cypripedium* comes from the Greek words for "Venus's slipper." Other common names also refer to the shape of the flower, and they include slipper plant and moccasin flower.

LILY, WOOD

Lilium philadelphicum

up to 3 feet tall

6 upward-facing petals
and sepals (tepals)

brown dots

1–3 red or orange
flowers

BLOOMS: Summer.

HABITAT AND RANGE: Wood lily is one of the more widespread lily species, occurring from British Columbia east to Quebec, south to Georgia, and west to Arizona. It prefers dry, open woods and woodland borders.

Columbian lily

Canadian lily

CONSERVATION: Wood lily is endangered or threatened in several states.

WILDLIFE PARTNERS: Wood lily is cross-pollinated by large butterflies such as swallowtails, monarchs, and the great spangled fritillary; and by hummingbirds, hummingbird moths, and bees. The larvae of smaller insects eat through the stem, often causing damage. Deer eat the leaves. Voles eat the bulbs.

IN THE GARDEN: There is great regional variation even within the same native lily species, so purchase bulbs from a source as local as possible. Lilies require well-drained soils that are rich in organic matter. Depending on the size of the bulb, they should be planted 4 to 8 inches deep. These are subject to damage from deer and other animals.

RELATED SPECIES:

Columbia lily (*Lilium columbianum*). Found in the West.
- nodding orange flowers
- dark spots
- tepals recurved
- 6 stamens

Canada lily (*L. canadense*). Found in the East.
- pale yellow to yellow flowers
- dark spots
- tepals arching outward or recurved

MEDICINAL USES: The Chippewa made a poultice from wood lily bulbs and placed them directly on sores, bruises, and wounds. The Iroquoi used tea made from the bulbs for stomach ailments, coughs, fevers, and to help women deliver the placenta after childbirth. The Dakota placed flower petals on spider bites.

The lily family is known to be over 50 million years old, and lilies have been cultivated since the Sumerian culture of the Tigris-Euphrates valley over 5,000 years ago.

Lilies were considered a sacred symbol of Venus, the goddess of love and beauty, and they were also associated with Juno, the goddess of marriage. White lilies have long been associated with the Virgin Mary, and they are considered a symbol of peace.

According to superstition, to dream of lilies during their growing season was thought to foretell marriage and happiness. According to the Victorian language of flowers, lilies are a symbol of majesty.

A major threat to all lilies, including the natives, is the scarlet lily beetle, which can do considerable damage. Hand-picking and destroying the adult beetles is a possible means of eradication.

The wood lily has upward-pointing blossoms, but because the anthers close their pores when it rains, the pollen is protected.

Scarlet lily beetle

MILKWEED, COMMON

Asclepias syriaca

2-6 feet tall

pinkish blossoms
in balls at the end
of stem

leaves opposite,
4 inches long

downy
underneath

unbranched stem

seeds are brown
on long silk

seedpods are
elongated wart

has both a corona
(crown-like structure)
and petals

BLOOMS: June–August.

HABITAT AND RANGE: Common in fields throughout the United States except in the Southwest and Washington, Idaho, and Wyoming.

CONSERVATION: Milkweeds are critical as a host plant for the health of the monarch butterfly population. Several milkweed species are federally threatened or endangered.

WILDLIFE PARTNERS: By far the most important wildlife partner for the milkweed is the monarch butterfly. Monarch populations have declined by 80 percent since the 1990s, largely due to the declining populations of milkweed. Planting regionally appropriate species will greatly help in conservation efforts. Milkweed also provides nectar and larval food for honeybees, bumblebees, wasps, and many other butterflies. The red milkweed beetle are common and found in profusion on many milkweed species.

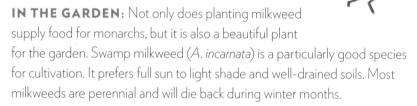

IN THE GARDEN: Not only does planting milkweed supply food for monarchs, but it is also a beautiful plant for the garden. Swamp milkweed (*A. incarnata*) is a particularly good species for cultivation. It prefers full sun to light shade and well-drained soils. Most milkweeds are perennial and will die back during winter months.

RELATED SPECIES: Butterfly weed (*Asclepias tuberosa*) features bright orange flowers that are borne on terminal clusters. The stem is rough and hairy, branching often toward the top. It is native throughout the East, and west to Texas, Colorado, and Minnesota. American Indians chewed on the stiff stalk of this plant, hoping to cure pleurisy. Because of this, it is sometimes known as pleurisy root. It was also (undeservedly) called chigger weed.

Numerous lance-shaped leaves are alternate and up to 6 inches long.

MEDICINAL USE: Milkweed was used by American Indian tribes for treating skin disorders such as ringworm, warts, and poison ivy. The indigenous peoples of the Quebec area used it as a contraceptive.

Seventy-three milkweed species are native to the United States. Monarchs use about one-third of these species as host plants. For more information about which species are useful and native to your region, go to monarchwatch .org. Southern milkweeds tend to be more toxic than their northern neighbors. Milkweed toxins can be harmful to livestock, pets, and people, but only in large quantities.

The genus is named for the Greek god of medicine, Asclepius, son of Apollo. The common name, milkweed, comes from the milky sap present in most species. This sap contains small amounts of rubber (about 2 percent) but not enough to make extraction worthwhile. Even so, during World War II, both the United States and Germany investigated the possibilities of using milkweed as a source for rubber.

The soft, downy seeds found inside the warty seedpod have proven to be effective insulators. They have been used as stuffing instead of feathers in beds and pillows, and even today milkweed is grown for the "down" that is used as a hypoallergenic filling for pillows. The downy seeds were at one time sewn onto woven cloth to give the appearance of fine, soft fur and were sometimes used in place of feathers on hats.

MONKEY FLOWER, LEWIS'S

Erythranthe lewisii (formerly known as Mimulus lewisii)

1-3 feet tall

pale pink (north)
to magenta (south)

dark lines and
maroon blotches

showy tubular
flowers

broad lance-shaped
opposite leaves

BLOOMS: June–September.

HABITAT AND RANGE: Common on stream banks and in moist meadows, usually at elevations between 4,000 and 10,000 feet, in Alaska, south through British Columbia to California, east to Utah and Colorado, and north to Montana.

CONSERVATION: Lewis's monkey flower (*Erythranthe lewisii*) is common and abundant. There are about 111 species in this genus, and 70 percent are native to California. Michigan monkey flower (*E. michiganensis*) is considered endangered within its range, as is the Carson Valley monkey flower (*E. carsonensis*) in California.

WILDLIFE PARTNERS: Monkey flower is an important plant for bees, which both collect the nectar and distribute the pollen. It is a host plant for the Baltimore checkerspot butterfly and the common buckeye butterfly, which lay their eggs on the foliage.

IN THE GARDEN: There are many species of perennial monkey flowers that do well in the garden and, when established, are drought tolerant. They grow in full sun to partial shade and look best when planted in groups.

MEDICINAL USE: Tea made from the monkey flower root was used in combination with other herbs by Iroquois women to treat seizures. Tribes from the Pacific Northwest used the petals to treat poisons and bee stings. A poultice from the mashed leaves and root was used for skin irritations and burns.

The *Mimulus* genus was reorganized in 2017, and most species were put into the *Erythranthe* genus. Some, particularly in the eastern United States, retained the name *Mimulus*, such as *Mimulus ringens*, Allegheny monkey flower. The genus name *Mimulus* and the common names all refer to the "faces" of the flower, which can appear to be smiling like monkeys.

The species, *Lewisii*, is named for Meriwether Lewis. It was originally collected in 1805 by the Lewis and Clark expedition.

American Indians and early European settlers ate the raw young leaves or cooked them with spinach or other greens. Also, because the plant absorbs sodium chloride from the soil, the ashes from burned leaves have a slightly salty taste and were used as a salt substitute.

PENSTEMON, PARRY'S BEARDTONGUE

Penstemon parryi

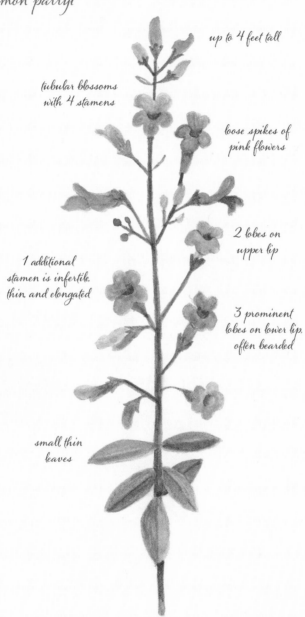

up to 4 feet tall

tubular blossoms with 4 stamens

loose spikes of pink flowers

2 lobes on upper lip

1 additional stamen is infertile, thin and elongated

3 prominent lobes on lower lip, often bearded

small thin leaves

BLOOMS: Late spring–early summer.

HABITAT AND RANGE: Not surprisingly, due to the size of the genus, penstemons grow in a wide variety of conditions, but they all seem to require well-drained soils. These are found from Alaska south to Guatemala. They are mostly a western genus, though a few species are found in the East.

CONSERVATION: Several species are threatened or endangered, including White River beardtongue (*Penstemon scariosus*)* and Graham's beardtongue (*P. grahamii*) in Utah (there are 71 species of penstemons native to Utah).

WILDLIFE PARTNERS: Penstemon foliage is important to deer and antelope. Birds eat the seeds. Penstemons tend to be prolific nectar producers and attract hummingbirds (particularly red-flowered species), butterflies, bees, and moths (particularly white-flowered species). Penstemon is particularly important to pollinators because it provides nectar when most spring flowers have finished blooming and summer flowers have not yet begun. The Yukon beardtongue (*P. gormanii*) is a host plant for the anicia checkerspot butterfly. The threatened Sacramento Mountains checkerspot butterfly is entirely dependent on the New Mexico beardtongue (*P. neomexicanus*). Protecting the penstemon resulted in the rebound of the associated butterfly.

IN THE GARDEN: Penstemons make excellent garden plants, and many hybrids have been developed from them. They are cold and drought tolerant and can grow throughout the United States. They have been highly hybridized (there are over 800 cultivars), especially in Europe.

MEDICINAL USES: The Navajo used penstemons to treat snake bites, vomiting, and wounds from guns and arrows. They believed it particularly valuable for treating eagle bites. The Blackfoot used it for stomach ailments.

---

* *Penstemon* spp. (the following information refers to North American penstemons in general).

American Indians used penstemons for cultural, medicinal, and culinary reasons. A delicious tea was made from the leaves and flowers of yellow penstemon. A blue dye made from some species was used by the Lakota to paint arrows, moccasins, and spears. The Apache believed that the plant was magical medicine. The Hopi associated it with the east (compass direction) in ceremonies, and, along with mint and sage; the Navajo used it in their shooting chant. The Zuni chewed the root of red penstemon and rubbed it over a rabbit stick (throwing stick) to bring good luck in hunting rabbits.

The common and genus name is from the Latin for five (*penta*) and the word for stamen, indicating the five stamens present on the flowers. The fifth stamen is prominent and often elongated and infertile. It usually protrudes from the petals, looking like a hairy tongue, giving rise to the other common name, beardtongue.

PHLOX

Phlox paniculata

Eastern swallowtail butterfly

2–4 feet tall

2–3 feet wide

tubular pink magenta or white flowers in domed clusters

dark green leaves, 4–6 inches

BLOOMS: July–September.

HABITAT AND RANGE: Phlox (*P. paniculata*) is found in open woods, along trails, and in thickets. It ranges from Iowa east to New York, and south to Georgia, Mississippi, and Arkansas.

CONSERVATION STATUS: Several phlox species are endangered or threatened, including the Texas trailing phlox (*P. nivalis* spp. *texensis*) and Yreka phlox (*P. hirsuta*), which is found only in Siskiyou County, California, in dry, open spaces.

WILDLIFE PARTNERS: Phlox are important pollinator plants. The blossoms are visited by hummingbirds and butterflies. The leaves are eaten by pronghorns (American antelopes), eastern white-tailed deer, and mule deer. Downy (or prairie) phlox (*P. pilosa*) is a critical host plant for the endangered phlox moth.

IN THE GARDEN: The greatest challenge in growing phlox is mildew. Avoid getting moisture on the leaves. Phlox has become a favorite of horticulturists. It is an easy-to-grow perennial. Selective hybridization has resulted in cultivars that are predictably white, pale, or dark pink.

RELATED SPECIES: Rocky Mountain phlox (*Phlox multiflora*) is found throughout the West and into Texas. It is low growing with abundant bloom in spring and summer. Drummond's phlox (*P. drummondii*) is an annual phlox that is beloved in the garden. It grows easily from seed. Woodland phlox (*P. divaricata*) also makes an excellent garden plant. It has light blue, violet, or pinkish flowers, is low growing at only 10 to 20 inches high, is lightly fragrant, and is semi-evergreen in warm climates.

MEDICINAL USES: Several American Indian tribes used the plant medicinally for a variety of purposes. Crushed leaves were made into a tincture for skin disorders, abdominal pain, and eye problems. Made into a tea, it was used as a gentle laxative.

Phlox blossoms were often included in bouquets, or tussie mussies (small bouquets
of flowers), during the Victorian era. Phlox
symbolizes a proposal of love and a wish for
sweet dreams.

The word *phlox* is from a Greek word meaning "flame" and refers to the
bright pink or red blossoms of some species.

There are about 67 species of the genus *Phlox*. One is native to Siberia,
and the remainder are native to North America, in habitats as diverse as alpine
meadows and deserts.

PITCHER PLANT, PURPLE

Sarracenia purpurea

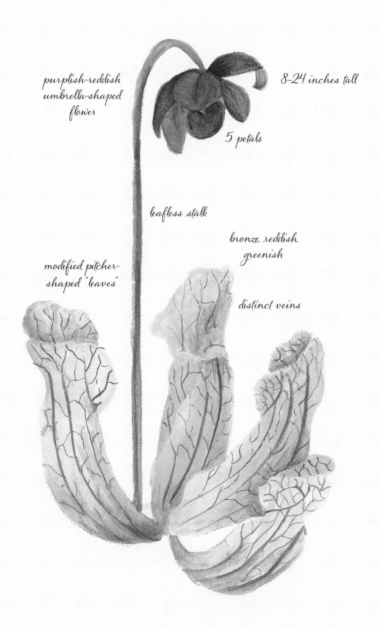

purplish-reddish
umbrella-shaped
flower

8-24 inches tall

5 petals

leafless stalk

bronze, reddish,
greenish

distinct veins

modified pitcher-
shaped "leaves"

GENERAL DESCRIPTION: There are 8 to 11 species of *Sarracenia* in North America. *Sarracenia* species are carnivorous plants that attract insects with color, nectar, and scent. Specialized leaves are "pitcher" shaped with a downward point and slippery hairs that lead to a wide cavity full of acids and enzymes. The insects can't crawl out of this, eventually die, and are digested by the plant. Most species of pitcher plant (but not *S. purpurea*) have developed a hood over the leaf that prevents excess rainwater from collecting and diluting the enzymes. Most *Sarracenia* species grow in clumps from 6 to 36 inches across. The scented flowers look like umbrellas and are found on stalks from 6 to 24 inches tall. They can be red, purple, pink, yellow, or white. New growth appears in April or May. The plant grows during the summer and goes dormant in fall.

BLOOMS: April–May.

HABITAT AND RANGE: Pitcher plants grow in wetlands—swamps, marshes, near springs, bogs, and at lake edges. They grow best in areas that are seasonally flooded. Pitcher plants are found throughout the East, north into southern areas of Canada, and west to British Columbia. They are most abundant in the southeastern United States.

CONSERVATION: Like many wetland species, *Sarracenia* species are threatened by habitat loss. Scientists estimate that over 97 percent of the original natural habitat for this genus has been lost.

WILDLIFE PARTNERS: The lip of the leaf has nectaries that emit a scent that attracts insects. The walls of the pitcher are waxy and slippery, trapping insects that often fall into a pool of enzymes at the base. The pitcher plant consumes insects such as ants, bees, wasps, beetles, snails, and slugs. The plants are most often pollinated by bees. The larvae of the pitcher plant moth (*Exyra* sp.) use it as a food source.

IN THE GARDEN: Pitcher plants need at least five hours of sunshine. They prefer seasonally flooded areas with acidic (pH 3–5) sandy soils.

RELATED SPECIES: The California pitcher plant (*Darlingtonia californica*) is in a different genus but the same family of pitcher plants. It is found sporadically throughout the Pacific Northwest.

MEDICINAL USES: The Cree used this plant to treat venereal diseases. The Cree and Algonquian-speaking tribes used the leaves to make a tincture to treat fevers and chills and to aid in childbirth. Algonquian-speaking tribes also used it as a diuretic and to relieve back pain and used the root mixed with beaver kidneys to treat urinary tract infections.

Early European plant explorers first described pitcher plants in 1570. In 1811, Benjamin Smith Barton wrote an article for *Philadelphia Magazine* describing pitcher plants. He indicated that they were used as cups to carry water and that the plant was commonly called water brash in New Jersey.

The Chippewa used the pitcher-shaped leaves as a drinking cup, a container for cooking, and as a toy. They thought that the powdered plant, when sprinkled on someone, was an effective love medicine.

PRAIRIE SMOKE

Geum triflorum

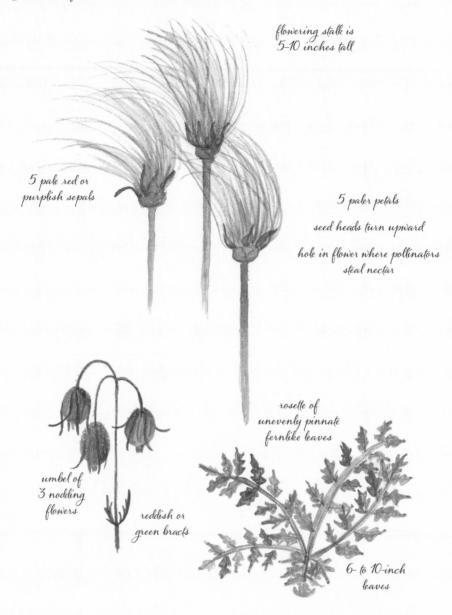

flowering stalk is
5–10 inches tall

5 pale red or
purplish sepals

5 paler petals

seed heads turn upward

hole in flower where pollinators
steal nectar

rosette of
unevenly pinnate
fernlike leaves

umbel of
3 nodding
flowers

reddish or
green bracts

6- to 10-inch
leaves

BLOOMS: April–May.

HABITAT AND RANGE: The plant is found in open prairies or partly shaded areas throughout the West, from British Columbia and Alberta, and south to Arizona and New Mexico. Also found in many northern states.

CONSERVATION: Abundant and common.

WILDLIFE PARTNERS: Bumblebees love this plant and gather its pollen by "buzz pollination," where the bumblebee grasps the flower and beats its wings until pollen shakes loose. Wasps, beetles, and small bees often cut holes at the top of the flower to steal the nectar. The foliage does not seem to be palatable to deer and other wildlife.

IN THE GARDEN: This makes a good garden plant for a partially shaded area with dry soils. The flowers are attractive for a long time period. This is commonly and enthusiastically included in western and northern gardens as a deer-resistant plant.

MEDICINAL USES: The Blackfoot used the crushed root along with other herbs in a sweat bath for rheumatism and stiff joints. They also made a tea drunk to help "build up the blood" and used it externally as an eye wash.

The Okanagan used this plant as a love potion. A woman who wanted to win back her man would trick him into eating the root of the plant. Mixed with tobacco, it was smoked to help clear the mind.

The seed heads are even more beautiful and conspicuous than the flowers. Long, airy, and feathery, they look like windswept hair. Another common name is old man's whiskers. Leaves turn beautiful orangey colors in fall.

The Chippewa chewed the dried root as a stimulant for endurance when long, arduous tasks were anticipated. The Paiute gave it to stimulate horses before a race.

The Blackfoot crushed the ripe seeds and used them as perfume.

ROSE, WILD
Rosa species (native species only)

2-6 feet tall

opposite leaves
with serrated
edges, 2-6 inches
long

5 pink
(rarely white)
petals

5 sepals

groups of
3-5 leaves

red fruits called
hips, hold 5-160
seeds

thorns are
really "prickles"

BLOOMS: Spring–early summer.

HABITAT AND RANGE: Sunny locations throughout North America, from Alaska to Baja, Quebec to Florida, and from the coast to the desert.

WILDLIFE PARTNERS: Native roses offer a wide variety of wildlife many benefits, including pollen, nectar, fruit, protection, and nesting sites. Rose flowers are important for many species of birds and butterflies. The wild prairie rose (*Rosa arkansana*) is thought to provide seeds for over 38 different species of birds. Many small mammals, including porcupines and deer, eat wild rose hips.

CONSERVATION: A few native rose species are considered rare or endangered in parts of their range, including the prickly rose (*R. acicularis*).

IN THE GARDEN: Native roses are not chemically dependent nor invasive enough to threaten the balance of a native ecosystem. Grown within their own region, many native roses are pest resistant and easy to grow. They need good, rich soil; sufficient irrigation; and a sunny spot.

MEDICINAL USES: Since Medieval times, roses have been used to treat a variety of ailments wherever they have grown. American Indians used the native roses to treat toothaches and earaches; diseases of the stomach, lungs, and intestines; overindulgence in wine; headaches; hemorrhages; sleeplessness; excessive perspiration; and hydrophobia.

Fossil records from the western United States show that the dog rose (*R. canina*) dates back 35 million years. The first cultivated roses were probably grown in Sumer, Babylonia, and Persia during the 3rd or 4th century B.C. Historical records indicate that during this same time period, the Chinese were cultivating roses to extract oil to make perfume.

Greek and Roman mythology is full of references to roses. Chloris, the Greek goddess of flowers, declared that the rose was queen of all flowers. The rose was dedicated to Harpocrates, the god of silence and secrets. The term *sub-rosa*, "under the rose," comes from the Roman practice of hanging a rose over a conference table. A code of honor dictated that nothing said under the rose would be repeated. Today sub-rosa still means secret or confidential.

Medieval gardens always included many roses, grown primarily as food, medicine, and for materials to make rosaries (created from compressed rose petals).

Rose hips from the dog rose (*R. canina*) are extremely high in vitamin C. During World War II, when imports of fresh citrus fruits were limited, rose hips were collected and eaten. The dog rose became a patriotic symbol in Britain. Today rose hips are used for making jams and jellies and tea.

Each rose color symbolized something different:

 WHITE: happiness and protection

 LIGHT PINK: grace and elegance

 DARK PINK: gratitude

 RED: passion and love

 YELLOW: joy

 ORANGE: enthusiasm

SHOOTING STAR

Dodecatheon pulchellum

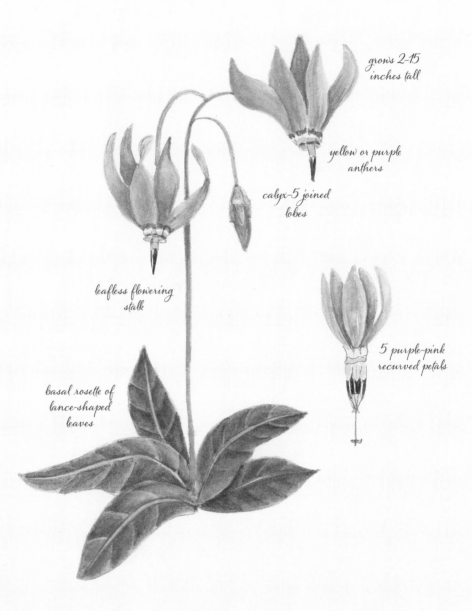

grows 2–15 inches tall

yellow or purple anthers

calyx–5 joined lobes

leafless flowering stalk

5 purple-pink recurved petals

basal rosette of lance-shaped leaves

BLOOMS: April–August.

HABITAT AND RANGE: The shooting star (*Dodecatheon pulchellum*) is one of the more widespread species of the genus. It can be found throughout the West, from Alaska south to Southern California, east to New Mexico, and north to the Dakotas.

CONSERVATION: Though most species are common and abundant in their range, a few species are threatened or endangered, including Sierra shooting star (*D. jeffreyi*) and jeweled throat shooting star (*D. radicatum*). Eastern shooting star (*D. meadia*), also called pride of Ohio, rooster heads, and prairie pointer, is listed as rare, protected, or endangered in many states.

WILDLIFE PARTNERS: Shooting stars are pollinated by bees who grab the petals and vibrate them by buzzing until the pollen is released. This is called "buzz pollination."

IN THE GARDEN: Shooting stars are commonly included in wildflower gardens, and they are considered easy to grow from seed. They need moist, well-drained, rich soils and protection from the sun (part shade or shade). *D. pulchellum* is hardy to 5°F.

One common name, prairie pointer, comes from the story that when the Western settlers saw these flowers growing, they believed that the point of the flower was pointing west for them. Another common name is pink darts. The flowers are said to smell like grape juice.

American Indian tribes made a tea from the leaf and used it to treat mouth cankers, especially in children. The tea was also used as eye drops.

SKYROCKET

Ipomopsis aggregata

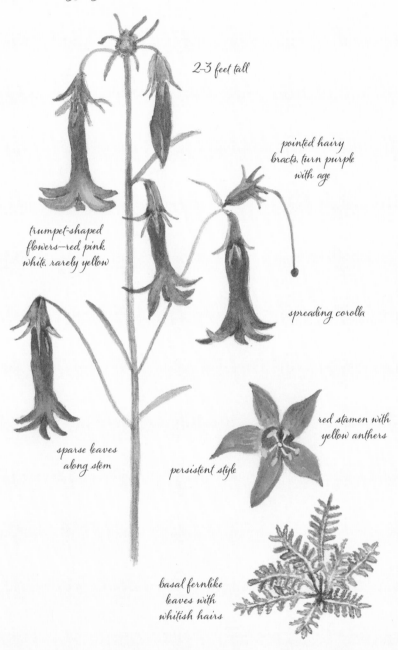

2–3 feet tall

pointed hairy bracts, turn purple with age

spreading corolla

trumpet-shaped flowers—red, pink, white, rarely yellow

red stamen with yellow anthers

sparse leaves along stem

persistent style

basal fernlike leaves with whitish hairs

BLOOMS: May–September.

CONSERVATION: Common and abundant.

HABITAT AND RANGE: Found throughout western North America from British Columbia to Mexico in a variety of habitats from dry slopes to open woodlands.

WILDLIFE PARTNERS: This plant is primarily pollinated by long-tongued moths and hummingbirds, bumblebees, solitary bees, and flies. The leaves and flowering stalks are eaten by elk and deer.

IN THE GARDEN: Easy to grow, this plant likes part shade and well-drained, sandy soils. Relatively short-lived, it dies after flowering. This is a biennial or short-lived perennial, and it forms a beautiful rosette of leaves the first year. After the plant flowers, it dies. However, if the flower stalk is eaten by wildlife before the plant has been pollinated, the plant will send up an additional stalk of flowers.

MEDICINAL USES: The Hopi used this for childbirth, and the Navajo used it to treat spider bites. The Zuni pounded the dried plant into a powder and applied it to the face for headaches and to help heal wounds.

Other common names for Skyrocket include scarlet gilia and scarlet trumpet. There are many different species of *Ipomopsis,* some of which show startling color combinations and beauty. The many-flowered gilia (*I. multiflora*), for example, has bright blue anthers.

Color variation in this plant is directly tied to its pollinators. At the beginning of the season, when hummingbirds are abundant, all the flowers are bright red or orange. As the season wanes and the hummingbirds begin to migrate, the flowers lose pigmentation and turn white, a color much preferred by the night pollinating hawk moth, which then begins to pollinate the flower. In warm or coastal regions, where hummingbirds do not hibernate, the flowers remain red throughout the season.

This plant was used by the Hopi as a dye, by the Ute to make glue, by the Navajo as a good luck charm for hunters and as ceremonial medicine. It was first collected in Idaho on the Lewis and Clark expedition in 1806.

TRILLIUM
Trillium species

Catesby trillium (*T. catesbaei*): strongly recurved petals, pink or white flower or short stem

Yellow trillium (*T. luteum*): yellow flower, sessile, no flower stem

Painted trillium (*T. undulatum*): flowers have single white petals with strong pink veins

Prairie trillium (*T. recurvatum*): purplish to brown flower, leaves mottled with purple, 3 reflexed sepals, grows 12–18 inches tall

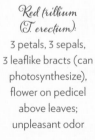

Red trillium (*T. erectum*): 3 petals, 3 sepals, 3 leaflike bracts (can photosynthesize), flower on pedicel above leaves; unpleasant odor

Toadshade (*T. sessile*): stalkless flower, maroon or brown, sometimes yellow, tightly held, looks almost closed; grows almost straight up from the stalk; "leaves" are mottled and stalkless

BLOOMS: Most trilliums bloom in early spring.

GENERAL DESCRIPTION: This genus is in the lily family, and it has some fairly distinctive characteristics. Most trilliums have three petals, three sepals, and six stamens. There are two different kinds of trilliums: sessile (meaning the flower is stalkless) and pedicellate (meaning the flower is on a short, usually nodding stem). There are no true aboveground leaves in trilliums. Instead, leaf-like bracts are prominent and can photosynthesize. Trillium flowers occur singly.

HABITAT AND RANGE: Trilliums are relatively common in the eastern United States and in the Pacific Northwest. Most species prefer rich woods, particularly deciduous woods, where they will receive plenty of light early in the season. Some like drier conditions.

CONSERVATION: Two trillium species are on the US Endangered Species list: relict trillium (*T. reliquum*), which grows in the Southeast, and persistent trillium (*T. persistens*), which was discovered in Georgia's Tallulah Gorge in 1950.

WILDLIFE PARTNERS: Several trillium species, in particular red trillium (*T. erectum*) and toadshade (*T. sessile*), do not have nectar but are pollinated by flies and beetles that are attracted to the scent of the flower, which most humans find fairly unpleasant. Trillium fruit looks something like a berry and is full of seeds that are covered with oily substances called elaiosomes (see Bloodroot, page 76). Ants and some wasps eat this substance and successfully distribute the seeds at the same time.

IN THE GARDEN: Most trilliums are difficult to establish in a wildflower garden. *T. sessile* is an exception to this. It is considered easy to grow and will spread well once established.

MEDICINAL USES: *T. sessile* was used extensively by American Indians to cure a variety of ailments. An eye wash was made from the sap of the root. The roots were also used to ease the pain of childbirth, a practice so common that *T. erectum* and western trillium (*T. ovatum*) were commonly known as bethroot, or birthroot. Trilliums were mixed with black cohosh (*Actaea racemosa*) to promote labor. They were also used to treat respiratory problems, and as a poultice it was used to treat ulcers and insect bites.

Across the globe, there are 50 species of trilliums, 39 of which are native to North America.

Botanical names are descriptive of the configuration of different species. *Erectum* refers to the erect posture of the petals of the red trillium, *Luteum* means yellow. *Undulatum* is from the Latin for wavy and refers to the undulations or crinkled edges of the petals of this species, and *recurvatum* refers to the recurved petals of the prairie trillium. *Catesbaei* is named for Mark Catesby (1683–1749), a British naturalist who discovered many plants in North America and sent specimens back to Europe.

Women commonly used trilliums as a love potion. They boiled the root then dropped it in the food of the desired man. An old American Indian story tells of a beautiful young girl who desired the chief's son for her husband. She boiled the root of trillium and was walking to put it in his food when she tripped. The root fell into the plate of an ugly old man who ate it. Unfortunately (for the girl, not the old man!), the love potion worked and he promptly fell in love.

An old mountain superstition says if you pick trillium, you will cause it to rain. A current saying about the trillium is that if you pick it, you'll kill the trillium, and this is absolutely true. The plant depends on the leaves to make food for the next year, and if the leaflike bracts are picked, then the plant more often than not will die. It is illegal to pick trilliums in Michigan and Minnesota.

White trilliums are the official emblem of the Canadian province of Ontario. Unfortunately, the plants are often decimated by an overenthusiastic deer population that eat them down to the nubs.

Trillium is a symbol of modest beauty.

VETCH, AMERICAN
Vicia americana

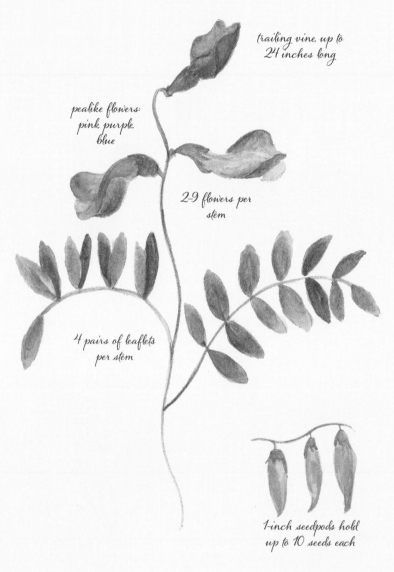

trailing vine, up to 24 inches long

pealike flowers: pink, purple, blue

2–9 flowers per stem

4 pairs of leaflets per stem

1-inch seedpods hold up to 10 seeds each

BLOOMS: May–August.

HABITAT AND RANGE: Found in open areas, borders, canyons, clearings, and road banks throughout the West and throughout the Northeast.

CONSERVATION: American vetch is generally abundant, though scarcer in parts of its natural range. It is considered endangered in Maryland. This plant can become aggressive and somewhat of a problem in some areas.

WILDLIFE PARTNERS: American vetch is an important and welcomed plant for many wildlife species. It provides habitat and food for game birds and small mammals and is eaten by deer and by both black and grizzly bears.

IN THE GARDEN: This plant makes a nice native alternative to crown vetch (*Coronilla varia*), which is often grown as a cover crop. American vetch needs full sun. It can become invasive, so care and control are necessary.

MEDICINAL USES: The Keres people of the American Southwest used the leaves to make a poultice used to treat spider bites and as an eye wash. The Okanagan used it to make a decoction used in a sweathouse.

Like other members of the pea family, *Vicia* can "fix nitrogen" in the soil. It is actually a bacteria (*Rhizobium*) that exists on the roots of vetch and other legumes that draws nitrogen gas from the air, converts it, and stores it in nodules found on the roots of the plants. When the plants die and 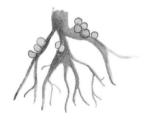 the roots decompose, they release nitrogen into the soil. Vetch makes a beautiful fall and winter ground cover that can be tilled under in the spring, allowing the roots to decompose and offer their gift of nitrogen.

Several American Indian tribes used the pod, seeds, and leaves as food. The fibrous roots were used as string.

The Yuki used the plant as a panacea or as a good luck charm. To keep a small bunch of the roots in your pocket was thought to bring good luck in gambling. The Navajo believed that smoke from the burning plants would improve the endurance of a horse before a race.

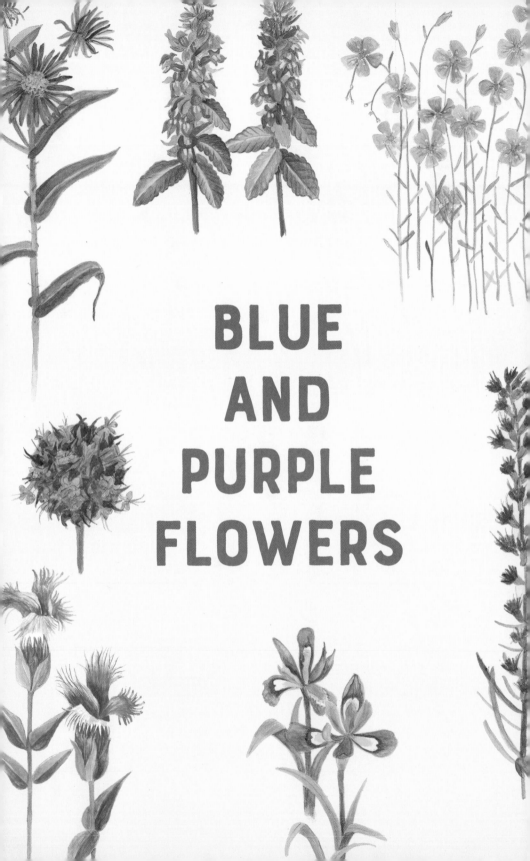

BLUE
AND
PURPLE
FLOWERS

Aster, New England *Symphyotrichum novae-angliae*

Blazing star *Liatris spicata*

Blue wild indigo, *Baptisia australis*

Camas *Camassia quamash*

Flax, wild blue *Linum lewisii*

Fleabane, showy (Aspen) *Erigeron speciosus*

Forget-me-not, alpine *Myosotis asiatica*

Gentian, fringed *Gentianopsis crinita*

Harebell *Campanula rotundifolia*

Iris, dwarf crested *Iris cristata*

Ironweed, New York *Vernonia noveboracensis*

Larkspur, Nuttall's *Delphinium nuttallianum*

Lobelia, great blue *Lobelia siphilitica*

Lupine, bigleaf *Lupinus polyphyllus*

Mint, wild *Mentha arvensis*

Monkshood, western *Aconitum columbianum*

Passionflower *Passiflora incarnata*

Salvia *Salvia* species

Violets *Viola* species

ASTER, NEW ENGLAND

Symphyotrichum novae-angliae

leaves become
smaller the higher
up the stem

4 feet tall

composite flower
with golden disc
flowers and purple
or pink ray flowers

basal leaves
1-5 inches
long

leaves clasp
the stem

slightly hairy
stem

BLOOMS: Late summer–fall.

HABITAT AND RANGE: Found in fields and open spaces. Widespread throughout the United States.

CONSERVATION: At least two species are known to be threatened: Georgia aster (*Symphyotrichum georgianum*) and serpentine aster (*S. depauperatum*).

WILDLIFE PARTNERS: Asters provide food and nectar for butterflies, bees, beetles, wasps, and flies. The larvae of the silvery checkerspot butterfly feed on the leaves. Deer and rabbits have been known to eat the tender new growth.

IN THE GARDEN: Asters are a great fall garden perennial. They need full sun and prefer rich, loamy soils. They can be planted from seed or small nursery plants. Provide plenty of moisture until they become established.

MEDICINAL USES: Smoke from burned aster leaves and flowers were inhaled by several American Indian tribes to treat mental illness, nosebleeds, headaches, and congestion and was sometimes included with other herbs for smudging and in sweat lodges. The Cherokee made a tea from the dried or fresh leaves for treating fevers, earaches, headaches, and dizziness. Currently, there is interest in the possibility of this plant's potential for soothing skin rashes caused by poison ivy and poison sumac. American Shakers used asters to make a salve to improve their skin. In Germany, a tea was drunk to cure the bite of a rabid dog. In traditional Chinese medicine, various aster species were used to relieve chronic coughs and clear phlegm.

Historically, aster species were found in the *Aster* genus. More recently, (primarily North American) native aster species are found in the *Symphyotrichum* genus.

According to Greek legend, asters were created from stardust when Virgo looked down from heaven and wept. Where her tears fell, stars began to sparkle. The word *aster* is from the Greek and Latin words for "star." Asters were considered sacred to all gods and goddesses, and wreaths made from aster blossoms were placed in temples during celebrations. Considered the herb of Venus, they were carried as talismans of love. Ancient Greeks believed that asters, strewn on the floor of a house, kept away serpents. The leaves boiled in wine and placed close to a hive of bee was said to improve the taste of honey.

In England, asters are called Michaelmas daisies because they are almost always blooming on St. Michael's Day, September 29.

In France, asters were thought to hold magical powers and were burned to keep away evil spirits.

BLAZING STAR

Liatris spicata

1-6 feet tall

blossoms lack ray flowers, only have pinkish-purple disc flowers

long white stamen make it look feathery

leaves become smaller up the stems

long, linear basal leaves

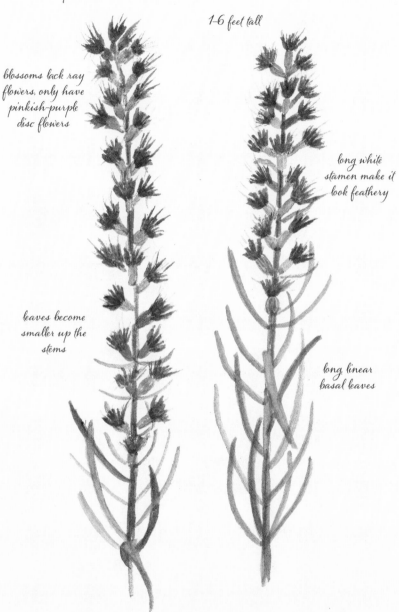

BLOOMS: July – September.

HABITAT AND RANGE: Found in fields and sunny, open places from the Midwest east to eastern Canada, through New England, and south to Georgia and northern Florida.

CONSERVATION: The threatened Heller's blazing star (*L. helleri*) is native to high elevations in the Southeast.

WILDLIFE PARTNERS: Flowers attract butterflies (sulphurs, swallowtails, monarch, Aphrodite fritillary, red admiral, painted lady, and more), bees (bumble, digger, and long-horned), and hummingbirds. Caterpillars such as those of the flower moths (*Schinia* spp.) feed on the flowers and seeds. The leaves are eaten by small mammals such as groundhogs, rabbits, and voles.

IN THE GARDEN: Full sun is necessary. Insufficient sunlight results in disappointing flowering. It prefers rich, well-drained soils. It provides excellent fall color.

RELATED SPECIES: Many *Liatris* species grow throughout the country. Dotted gayfeather (*L. punctata*) is very similar and is found in the plains of the Midwest and west to the eastern edge of the Rocky Mountains.

MEDICINAL USE: Some American Indian tribes used various *Liatris* species as a diuretic, stimulant, and expectorant. The Omaha tribe used it for abdominal pain and muscular spasms. The Cherokee used it for back and joint pain.

The name blazing star was given because the bright magenta blossoms are grouped together like stars. Another name, gayfeather, is from the feathery look of the densely packed blossoms.

This makes a great cut flower and is used in potpourris or as an insect repellent.

BLUE WILD INDIGO

Baptisia australis

TOXIC

2-4 feet tall

bilaterally symmetrical flowers

1-inch flowers clustered along flowering stalk

5 joined sepals to make cup or tube

female ruby-throated hummingbird

leaves divided into 3 leaflets

light to deep blue or purple flowers

seedpods 2½-3 inches long, turn black when ripe

BLOOMS: April–June.

HABITAT AND RANGE: This prairie plant occurs in open fields, prairies, and forest borders. It is also found in sandy, well-drained open areas throughout central and eastern United States and is naturalized in many places.

CONSERVATION: This species is considered threatened in Maryland and Indiana, and it is endangered in Ohio, primarily due to habitat loss. It is an important host plant for the rare frosted elfin butterfly. There are several other *Baptisia* species that are endangered or of concern.

WILDLIFE PARTNERS: Because the plant is toxic and bitter, it is seldom eaten by animals. In spite of this, it is a host plant for the larvae of several butterflies, including the orange sulfur and clouded sulfur, frosted elfin, eastern tailed blue, and wild indigo duskywing. It is pollinated by native bees, bumblebees, and other insects.

IN THE GARDEN: Baptisia is a very good garden plant. It tolerates a wide range of soil conditions but prefers moist, well-drained soil. It needs at least six hours of sunlight. The leaves turn a beautiful silver in fall. This plant was named the 2010 Perennial Plant of the Year by the Perennial Plant Association.

MEDICINAL USES: The Osage made a tea from leaves as an eyewash. Several tribes used the tea to stop vomiting, ease pain from toothaches, and as a general tonic. It was at one time used by early settlers to treat the flu, colds, malaria, and typhoid but is no longer considered safe to take by mouth or put on skin. *Toxic.*

Though beautiful, these flowers are completely unsatisfactory as a cut flower as they wilt as soon as they are picked. The sap in the stem turns a slate blue color when exposed to the air. The long seedpods, which were used by American Indian children as rattles, are attractive and sometimes used in flower arrangements.

Both the common and the genus name come from its use as a dye. *Bapto* means to dye, and the roots were used by the early European settlers as a poor substitute for true indigo. The Cherokee extracted a blue dye from the roots and used it to dye cloth.

CAMAS

Camassia quamash

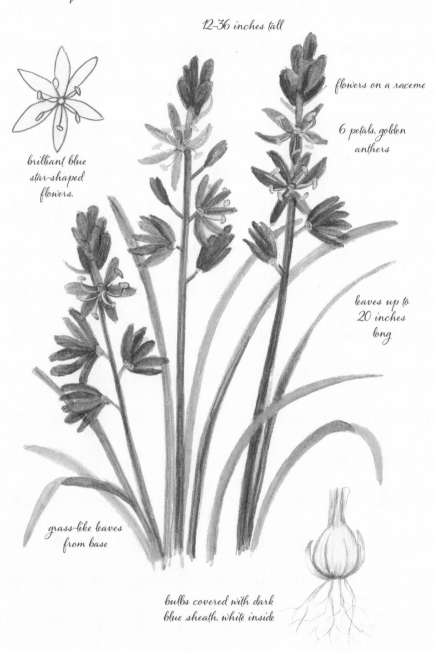

12-36 inches tall

flowers on a raceme

6 petals, golden anthers

brilliant blue star-shaped flowers,

leaves up to 20 inches long

grass-like leaves from base

bulbs covered with dark blue sheath, white inside

BLOOMS: April–June.

HABITAT AND RANGE: Found in moist fields, some years covering an entire meadow. Grows throughout the northwestern United States and British Columbia, south to California, and east to Wyoming and Montana.

CONSERVATION: Atlantic camas (*Camassia scilloides*) is found throughout the eastern part of North America and is considered threatened or endangered in at least four New England states.

WILDLIFE PARTNERS: Camas is pollinated by native bees, and the foliage is eaten by elk, deer, and moose. Pocket gophers store the bulbs in their holes in huge numbers.

IN THE GARDEN: Several cultivars have been developed from the wild species. They need well-drained, loamy soils. Dig and divide in fall after the leaves begin to dry and wither.

This was one of the most important wild foods for many American Indian tribes, including the Nez Perce, Cree, Blackfoot, Flathead, and Coastal Salish. The camas bulb caches of the pocket gophers were often robbed by Nez Perce women. In addition, the camas root provided the Lewis and Clark expedition with a much-needed food source. Western settlers rarely used it, but their presence greatly impacted the American Indian usage of camas root. As the

From bottom up: Ponderosa pine branches, river boulders, fire, balsam and skunk cabbage leaves, camas bulbs, Douglas fir tips, bark, dirt, fire

settlers continued their Western expansion, both grazing cattle and the plow destroyed much of the camas root habitat. This contributed to rising tensions between the settlers and indigenous peoples.

For the tribes that dug and preserved the roots, camas was an essential food source. Primarily women dug and baked the bulbs. When there was time, great pits were dug for roasting the bulbs.

It was necessary to roast the bulbs before eating. It was said to taste something like sweet potato, but sweeter. The Nez Perce name for this root actually means "sweet." The baked roots were also ground into flour to make small cakes, used in making gravy, and boiled to make a beverage something like coffee. Tea made from camas roots and mixed with honey was sometimes used as a cough syrup.

The baked bulbs could be preserved for years and were, in some areas, an important trade item. It was so valued that small bags of dried camas were given as gifts at weddings and funerals.

Many tribes celebrated camas with ceremony. The Camas Dance was often held twice a year. The first was in January, the second in summer, immediately after the bulbs had been harvested.

Note: The so-called Death Camas is a white flowering bulb of a completely different genus, *Toxicoscordion,* and it is *extremely toxic.* It blooms at the same time that blue camas do. Although the flowers do not look alike, the bulbs are almost identical.

FLAX, WILD BLUE

Linum lewisii

18-20 inches tall

bright blue flowers

wedge-shaped petals ¾-1½ inches

dark veins are nectar guide

stems very leafy when young

linear leaves

slender stems

5 sepals,
5 white stamens
5 styles

BLOOMS: March–September.

HABITAT AND RANGE: Ridges and dry slopes, prairies, and meadows from sea level to elevations of 11,000 feet in the western three-quarters of North America.

CONSERVATION: Abundant, can be aggressive. The European species, common flax (*Linum usitatissimum*), often escapes from cultivation and becomes invasive.

WILDLIFE PARTNERS: Flax is pollinated by native bees and flies. It is hermaphroditic (both sexes are on a single plant) and can be wind pollinated, but insects are needed for seed production. Birds eat the seeds in winter. It provides forage material for deer and antelope and cover for small birds.

IN THE GARDEN: This plant is easy to grow, comes quickly from seed, and is drought tolerant. It needs a sunny, dry spot but can become an aggressive grower and may be slightly invasive.

MEDICINAL USE: The Shoshone used a tincture made from the root as an eye medicine. The Navajo used it for headaches and as a fumigant. Various tribes used a poultice made from the entire plant on bruises and swelling.

Nine species of this genus are native to North America. The European species, common flax (*L. usitatissimum*), which means "most useful," is an important commercial plant. It is the source of fiber for making linen. In fact, the word *linen* comes from the genus name *Linum*. The history of linen is ancient, dating back possibly 10,000 years to the lake people of Switzerland. Egyptians used linen extensively for making garments and wrapping mummies. Linen has been used for such diverse purposes as printed currency and for making sailcloth. In addition, flaxseeds are used for eating and cooking and for making linseed oil.

American Indians used the plant for making cordage. The Klamath, Plains, and Great Basin tribes made the cords into mats or into mesh for making snowshoes and fishing nets. The Omaha and other tribes used the seeds as flavoring in cooking. Okanagan young women made a concoction from the upper parts of the plant (stems, leaves, and flowers) to use as a body and hair wash.

L. lewisii is named for Meriwether Lewis.

FLEABANE, SHOWY (ASPEN)

Erigeron speciosus

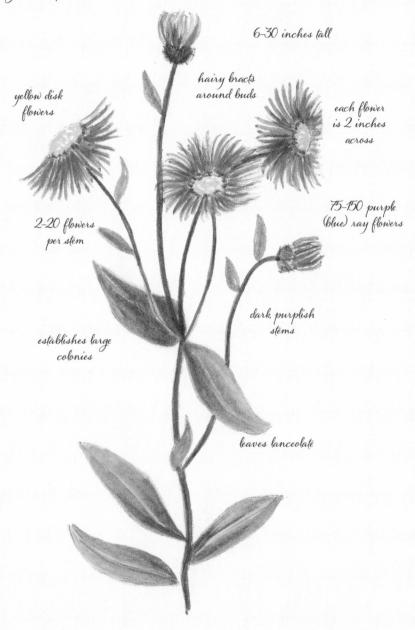

6–30 inches tall

hairy bracts
around buds

yellow disk
flowers

each flower
is 2 inches
across

2–20 flowers
per stem

75–150 purple
(blue) ray flowers

dark purplish
stems

establishes large
colonies

leaves lanceolate

BLOOMS: June–September.

HABITAT AND RANGE: Showy fleabane is found in sunny meadows and mountain slopes throughout the west from Alberta and British Columbia throughout the Mountain states, and south to Arizona and New Mexico.

CONSERVATION: Several native species are threatened or endangered.

WILDLIFE PARTNERS: *Erigeron* species attract quite a variety of insects, including bees, wasps, flies, beetles, and small butterflies. Larvae of many species eat the leaves and roots. Deer and groundhogs eat the leaves and mice eat the seeds.

IN THE GARDEN: This plant has been hybridized to produce some really wonderful, long-blooming cultivars with names such as 'Pink Jewel' and 'Azure Fairy.' They need very well-drained soil and full sun.

RELATED SPECIES: Daisy fleabane (*Erigeron annuus*) can grow 3 to 4 feet in height; the stem is branched, and the leaves and stem are covered with tiny hairs. The flowers are small and daisylike and are purple or white, sometimes tinged with pink. There are usually several flowers to a stem. It is a very common plant, native to 43 of the lower contiguous states.

There are 173 species of Erigeron that are native to North America. The genus name comes from two Greek words that mean "early" and "old man." Some people say the name comes from the thin petals that look like an old man's beard; others say it is because many species bloom early in the year. The species name, *speciosus,* means "pretty."

Early Western settlers used dried fleabane for pillows and mattresses, with the hope that it would keep away fleas and unwanted insects, which unfortunately, it did not.

The old English name for this is poor robin's plantain because the seeds were said to have been exported to Europe inside a stuffed robin.

An old superstition says that if a pregnant woman wants to know the sex of her baby, she should plant fleabane. When the flowers bloom, if they are tinged with pink, she'll have a girl; if with blue, a boy.

FORGET-ME-NOT, ALPINE

Myosotis asiatica (also seen as Myosotis alpestris)

up to 16 inches tall

5 petals from
short tube

sky blue flowers
with yellow centers

flowers grow close
to the stem

leaves and stems
have stiff hairs

many stems
produce rounded
look

leaves are
oblong and
alternate

BLOOMS: June–September.

HABITAT AND RANGE: The alpine forget-me-not (*M. asiatica*) grows on hillsides and meadows at elevations from 5,000–16,000 feet. This plant grows in Alaska, Canada, Colorado, Idaho, Montana, Oregon, Washington, and Wyoming. As the species name suggests, this plant is native not only to North America but to Asia, Northern Europe, and Russia.

CONSERVATION: The water forget-me-not (*M. scorpioides*), which is a European species, can be aggressive and invasive.

WILDLIFE PARTNERS: This plant is pollinated by bees, butterflies, flies, and moths. The larvae of several moth and butterfly species use this as a host plant.

IN THE GARDEN: Seeds are available commercially, and it is fairly easy to grow within its range, preferring dry soils.

RELATED SPECIES: The small forget-me-not (*Myosotis laxa*) has flowers ¼–⅓ inch across. They are found on a coiled stem like a scorpion tail, thus the other common name, scorpion weed.

MEDICINAL USES: The forget-me-not has been used for centuries in Europe as medicine for treating lung ailments and nosebleeds. The leaves, boiled in wine, were used as an antidote for an adder's bite. Based on the doctrine of signatures, which suggested that a plant's physical appearance indicated its medicinal use, early colonists used forget-me-not to treat spider and scorpion bites.

This is the state flower of Alaska, designated in 1917. The symbolism for this flower is constancy and perseverance, the same characteristics shown by the intrepid early settlers in Alaska. In Europe, forget-me-nots were associated with loving remembrance, friendship, and fidelity.

A German story explains the flower's name: A man and his lover were walking in a high mountain meadow when she saw this beautiful blue flower and asked for it. The man, reaching for the flower, lost his footing. He tossed the flower to the woman and cried out, "Forget me not!" as he fell to his death.

Egyptians believed that if you put the leaves over your eyes when you slept during September, the month of Thoth, you would have visions.

Forget-me-nots played a small role in English history. In 1398 King Richard II, feeling threatened by Henry of Lancaster, banned him from England. Henry and his followers adopted this blossom as their symbol, promising to return and take the throne. Eventually, he succeeded and became Henry IV.

The genus name means "mouse ears" because it was thought the oval young leaves of some species looked like mice ears.

GENTIAN, FRINGED

Gentianopsis crinita

grows 1-3 feet tall

petals are cut
in long fringes
at the tips

sky blue tubular
flower 2 inches
long

lanceolate leaves
with pointed tips,
rounded at base

flowers occur at
the stem ends

BLOOMS: Late August–November.

HABITAT AND RANGE: Fringed gentian (*G. crinita*) is found in moist meadows, stream banks, and woodland borders where there is sufficient moisture. It grows in the East from Canada, south to Georgia, and west to Illinois, Iowa, and North Dakota.

CONSERVATION: This plant is rare wherever it grows, and it is ranked as endangered in many states.

WILDLIFE PARTNERS: It's a great pollinator plant for bumblebees.

IN THE GARDEN: This is difficult to grow, and, as a biennial, it lasts only two years. It comes better from seeds rather than transplants.

RELATED SPECIES: The Sierra fringed gentian (*Gentianopsis holopetala*) is found in the Sierra Nevada mountains. The Rocky Mountain fringed gentian (*G. thermalis*) was so named because it grows in great abundance around the thermal pools in Yellowstone Park.

MEDICINAL USE: This plant was used against the plague and digestive ailments in ancient times. Leaves were made into a poultice to treat infections, cancer, and wounds. Together with elderflower, verbena, cowslip, and sorrel, fringed gentian was used to treat sinus infections. The Catawba made a compress of gentian leaves to relieve backaches.

Even though these blossoms are startlingly blue, they contain no blue pigment. Actually, blue pigment does not exist anywhere in the flower world. Instead, blue plants have varying amounts of anthocyanins (red pigments) that appear blue with varying pH levels.

The first mention of the medicinal value of gentian is in papyrus scrolls found in an Egyptian tomb at Thebes, dating back 3,000 years.

According to Pliny, the Roman naturalist, gentians were named for Gentius, king of Illyria. During the 2nd century B.C. Gentius was not only king but also a skilled healer.

A Hungarian legend says that gentians first grew centuries ago when a plague killed many people. A king, Ladislaus, felt powerless to help, and in desperation, he shot an arrow into the air, praying to God that the arrow would land on a plant that would help save his people. Legend says that the arrow landed on a gentian and that the plant was used to save many people.

Early Western settlers flavored gin or brandy with gentian, and they drank it as a spring tonic or to aid in digestion. In Europe, many commercial aperitifs contain extract of gentian.

Gentian root was carried as a good luck charm in southern Appalachia.

Gentian flower became the emblem of the Japanese Minamoto clan, one of the four great clans in Japan during the Heian period.

Gentian is one of the few wildflowers to have a poem written to and about it. In "To the Fringed Gentian" by William Cullen Bryant, this beautiful flower is "colored with the heaven's own blue."

HAREBELL

Campanula rotundifolia

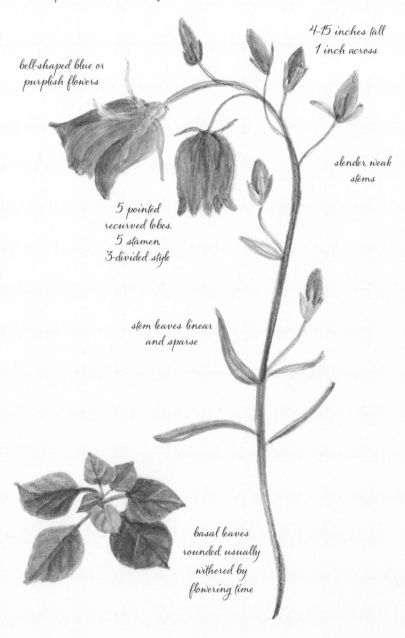

4–15 inches tall
1 inch across

bell-shaped blue or
purplish flowers

slender weak
stems

5 pointed
recurved lobes,
5 stamen,
3-divided style

stem leaves linear
and sparse

basal leaves
rounded usually
withered by
flowering time

BLOOMS: June–October.

HABITAT AND RANGE: Open meadows, boggy areas, or hillsides up to 12,000 feet. This plant is found throughout the West (except for Nevada) and in the East from New England south to Tennessee and Virginia.

CONSERVATION: At least one species, Brooksville bellflower (*Campanula robinsiae*), is on the Florida Endangered Species list. A Eurasian species, creeping bellflower (*C. rapunculoides*), is widely grown in gardens but has escaped and is now considered a noxious weed.

WILDLIFE PARTNERS: This plant attracts hummingbirds and is also pollinated by bees.

IN THE GARDEN: Native harebells are easy to grow and are a great plant for the garden if planted in clumps, similar to their growth habit in the wild or in a rock garden. They tolerate a variety of conditions, including full sun, part shade, or full shade. They like well-drained soils. The flowers last a long time.

Other names for this plant include bluebells, bluebell of Scotland, fairies' thimbles, and dead men's bells. The genus name, *Campanula,* is from the Latin for "little bell."

Harebells were at one time made into blue dyes used for making tartans in Scotland. This flower became the symbol of the MacDonald clan.

The Haida tribe of the Pacific Northwest called these flowers the "blue rain flowers" and believed that picking them would cause it to rain.

Harebells have been associated with witches and death. It was thought that if witches ate this plant, they would turn into hares that were reputed to bring bad luck to all who saw them. This accounts for the additional common names witches'-thimble and harebell.

The European bellflower rampion (*C. rapunculus*) was at one time widely grown in European gardens for its spinach-like leaves and for its parsnip-like root. It was made famous by the Brothers Grimm. In the story of Rapunzel, a man stole a piece of this bellflower for his pregnant wife who craved it. The garden was owned by a witch, who caught and threatened him. In order to get away, the man promised the witch that when the child was born, the baby would be given to her. The witch demanded the newborn and named her after the plant that her mother craved, Rapunzel. Rapunzel, of course, let down her hair later in life.

The bulbs from some *Campanula* species are used as glue for bookbinders or as a substitute for starch.

Bellflower is the symbol of constancy and kindness.

IRIS, DWARF CRESTED

Iris cristata

MILDLY TOXIC

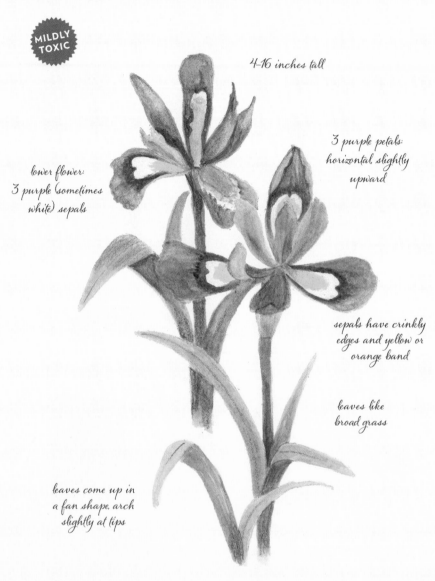

4–16 inches tall

3 purple petals: horizontal slightly upward

lower flower: 3 purple (sometimes white) sepals

sepals have crinkly edges and yellow or orange band

leaves like broad grass

leaves come up in a fan shape, arch slightly at tips

BLOOMS: March–May.

HABITAT AND RANGE: The dwarf crested iris (*I. cristata*) grows in upland woods and partly shaded forest edges primarily in the East from New York south to Florida and west to Arkansas, Missouri, and Ohio.

CONSERVATION: *I. cristata* is common and abundant. Dwarf lake iris (*I. lacustris*) looks very similar to dwarf crested iris, but it is an endangered species found only in Michigan, Wisconsin, and Ohio. Yellow flag iris (*I. pseudacorus*) is considered invasive in wetland areas, though it is sometimes used to help clean water at water purification plants.

WILDLIFE PARTNERS: Irises attract bees and hummingbirds. Mammals tend to avoid the foliage, perhaps because it is slightly toxic.

IN THE GARDEN: *Iris cristata* is a favorite for a native plant garden. It's easy to grow in well-drained, rich soils in partial shade. Best propagated by dividing the rhizomes in early spring or fall.

RELATED SPECIES: The Douglas iris (*Iris douglasiana*) is a common wildflower in the coastal regions of central California and north through the Pacific Northwest. It has light lavender to almost-white blossoms, grows 18 inches tall, and forms clumps that measure about 3 feet across.

MEDICINAL USE: American Indians used the rhizome to make a poultice to treat sores and wounds. It was also used internally as a cathartic. *Mildly toxic.*

Approximately 28 iris species are native to the United States, and they are found in every state except Hawaii.

The genus and common name were named by the Greeks for the goddess of the rainbow, Iris, because it was said that she wore robes of many colors, just as the iris blooms in many colors. Since one of the duties of Iris was to lead the souls of women to the Elysian Fields after they died, the Greeks often put iris blossoms on the graves of their women. The word *iris* means "eye of heaven" and was given both to the center of the eye and to the goddess.

France has used the iris as a symbol of their victories since the 1st century. Legend tells us that after an important victory, the king of the Franks, Clovis I, changed the symbols on his banner from three toads to the yellow flag iris. Louis VII revived the iris symbol and used it as the basis of the fleur-de-lis (flower of Louis).

Two European irises, the German iris (*I. germanica*) and sweet iris (*I. pallida*), are cultivated for use in the perfume industry. The rhizomes of these two species combined are called orris root, and they are used as a stabilizer and fixative.

Many different kinds of gin are flavored and colored with irises.

Yellow flag iris yields roots that are useful for making brown and black dye and black ink. Western species with tough, fibrous leaves, such as *Iris douglasiana*, were used by the Yuki to make cordage and rope.

In China, iris was planted by a garden gate in the belief that it would keep snakes from entering the garden.

IRONWEED, NEW YORK

Vernonia noveboracensis

3-6 feet tall

clusters of
purple or
magenta flowers

each blossom has
30-50 disk flowers

serrated leaves,
narrow and
pointed at end

single stem
branches
toward top

BLOOMS: Late summer–fall.

HABITAT AND RANGE: There are 6 to 8 ironweeds native to the eastern United States. All are found in open spaces and woodland edges. *V. noveboracensis* is found from New England to the southeastern states, only as far west as Alabama, Kentucky, and West Virginia.

CONSERVATION: There are 1,000 species of *Vernonia* worldwide, several of which are endangered or threatened, including Proctor's ironweed (*V. proctorii*), endemic to Puerto Rico and on the US Endangered Species list.

WILDLIFE PARTNERS: This plant is an important pollinator for bees (particularly those of the *Melissodes* genus) and butterflies as it blooms late in the season. It is the larval host for the American painted lady butterfly. Songbirds such as goldfinch, house finch, and song sparrows eat the seeds.

IN THE GARDEN: New York ironweed is a great fall perennial. It tends to be tall and robust, so place it at the back of the garden. It likes full sun or part shade and moist, well-drained soils. Cutting it back in spring results in a bushier, more stout plant.

RELATED SPECIES: The related giant ironweed (*Vernonia gigantea*) has a slightly broader range, growing west as far as Texas and north to Kansas, Iowa, and Michigan. Though New York ironweed (*V. noveboracensis*) has more concentrated blooms, *V. gigantea* grows even taller, up to 7 feet.

MEDICINAL USE: The Cherokee made a tea from leaves to use as an aid in childbirth and as a general blood tonic. They also used the plant to treat ulcers, hemorrhaging, and loose teeth. Modern testing has found that *V. condensata*, a species native to Brazil, has leaves that may be a promising source of antioxidants.

The genus was named for an English botanist, William Vernon, who worked extensively at collecting and identifying plants in North America in the late 17th century. The common name is so named because the stem, or stalk, is so sturdy it's "like iron." This makes it an unwelcome plant for farmers who find it difficult to cut it down. Livestock will not eat it, so it sometimes spreads and becomes a pest.

The Yuchi of eastern Oklahoma (formerly located in the southeastern United States) uses smooth ironweed (*V. fasciculata*) as an important part of their annual Green Corn Festival, which is still celebrated today. The Festival traditionally occurs in July or August and is considered a New Year's celebration. Four boys are chosen by the chief to be "pole boys." The chief places the root of vernonia (which looks like a bear's paw) in the pocket of each boy chosen. These roots are used in the ceremony and then burned at the end.

Various species of ironweed found in West Africa are widely used as a food plant. Vernonia is one of the most widely eaten leaf plants in Nigeria.

"You belong among
the wildflowers . . .
You belong somewhere
you feel free . . ."
—Tom Petty and the
Heartbreakers, "Wildflowers"

LARKSPUR, NUTTALL'S

Delphinium nuttallianum

TOXIC

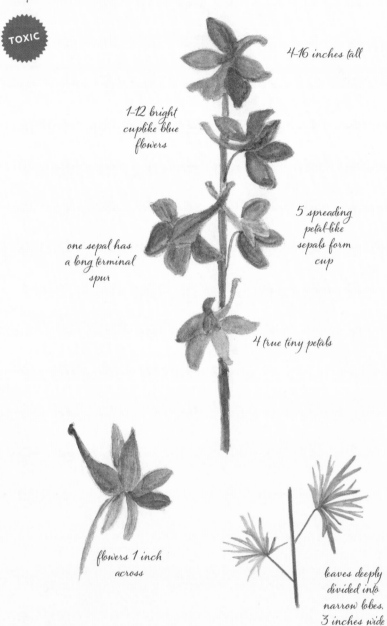

4–16 inches tall

1–12 bright cuplike blue flowers

5 spreading petal-like sepals form cup

one sepal has a long terminal spur

4 true tiny petals

flowers 1 inch across

leaves deeply divided into narrow lobes, 3 inches wide

BLOOMS: March–July.

HABITAT AND RANGE: Nuttall's larkspur (*D. nuttallianum*) is found in shrubland, desert, and meadow edges throughout the West.

CONSERVATION: Several species of larkspur are critically endangered and protected. Among these are Baker's larkspur (*D. bakeri*), Casey's larkspur (*D. caseyi*), tall larkspur (*D. exaltatum*), peacock larkspur (*D. pavonaceum*), and yellow larkspur (*D. luteum*).

WILDLIFE PARTNERS: Though most larkspurs are toxic to humans, wildlife, and livestock, it is used as a larval food by the dot moth. *D. nuttallianum* is pollinated both by queen bumblebees and hummingbirds, though its blue color makes it atypical for this.

IN THE GARDEN: Larkspurs make great garden plants. Low-growing species look best planted in groupings. They need full to medium sun and well-drained, rich soil. For the back of a mixed border, try tall larkspur (*D. exaltatum*), which is a species native to the East.

RELATED SPECIES: Carolina larkspur (*Delphinium carolinianum*) is found in open prairie or grassland from Virginia and Missouri south to Texas and Florida. This larkspur can cover acres of land in its native habitat.

There are over 300 species of *Delphinium* worldwide, and 60 species native to North America. The name *Delphinium* comes from the Greek word for dolphin, because the long spur prominent on so many species looks like the nose of a dolphin. The Spanish name for this is *espuela de caballero,* or the spur of

the horseman, again referring to the long spur of the flower. The name larkspur is from the resemblance to the crest of the bird, the lark.

All parts of this plant are toxic and can be dangerous or even fatal if ingested. The seeds, soaked in alcohol and ground into a paste, were used to kill head lice.

Nuttall's larkspur was used by the Okanagan to make dye or ink to color arrows, and it was used in various ceremonies by the Navajo. Tribes of the Pacific Northwest Coast used the blossoms of their local coastal larkspur (*D. decorum*), mixed with a glue made from salmon and native grapes to make paint for arrows and bows. The little larkspur (*D. bicolor*) was used by Blackfoot women to shine and straighten their hair.

Almost all larkspur species make great cut flowers.

LOBELIA, GREAT BLUE

Lobelia siphilitica

TOXIC

1-4 feet tall

upper flowers divided into 2 parts, bottom into 3 parts with white splotches

flowering stalks originate in leaf axils

calyx has 5 pointed lobes, covered with hairs

leaves are oval 2-6 inches long, slightly lobed with tiny teeth

bright blue bilaterally symmetrical 1-inch flowers

BLOOMS: August–September.

HABITAT AND RANGE: Great blue lobelia (*L. siphilitica*) is found abundantly in moist meadows, rich woodlands, and swamps from Maine south to Georgia, west to Texas, and north to North Dakota.

CONSERVATION: There are 43 species of *Lobelias* native to the United States, and their conservation status varies from locally abundant to critically endangered. Hawaii is home to several different species, including Oahu lobelia (*L. oahuensis*), which is found on cliffsides on the island of O'ahu, and *Lobelia monostachya,* which was once thought extinct but was rediscovered on O'ahu. Both are threatened by invasive plants, which compete for available space and resources, and by habitat destruction caused by feral pigs.

WILDLIFE PARTNERS: Great blue lobelia and the red cardinal flower are pollinated by bumblebees, many species of native bees, butterflies, and hummingbirds.

IN THE GARDEN: Lobelias make an excellent addition to the wildflower garden. Pinch back the great blue lobelia during the active growing season to encourage bushier plants and more blooms. The great blue lobelia prefers semi-shade and moist conditions.

RELATED SPECIES: Cardinal flower (*Lobelia cardinalis*) has brilliant red bilaterally symmetrical flowers that appear from July to September. It is found along stream banks and in moist places. In the garden, provide semi-shade and moist conditions.

MEDICINAL USES: American Indians and Europeans used the great blue lobelia as a treatment for syphilis. A tea from the roots was used to help cure intestinal worms. *Toxic.*

Eighteenth-century Europe was desperate for a cure for syphilis, which was both widespread and all too often fatal. It was no surprise that plant explorers in the New World were hoping to find medicinal plants that would cure this and other diseases. Sir William Johnson, superintendent of Indian Affairs in North America from 1756–1774, visited various tribes, hoping to find medicinal plants. When he heard that the Iroquois used great blue lobelia as a cure for syphilis, he acquired samples and sent them to England. Eventually, all tests for its efficacy came back negative, and so the search for a cure continued, but not before the plant had been named *Lobelia siphilitica*. British gardeners continued to grow lobelia in their gardens for its beauty.

Cherokee Indian women dug red cardinal flower roots from the ground with much ceremony, then touched it to all parts of their bodies to make them desirable. Many tribes believed that it was a potent love charm.

The common name, cardinal flower, came from the similarity in color between the flower and the scarlet robes worn by the cardinals in the church. This name was also given to the songbird.

According to the language of flowers, cardinal flower means distinction and splendor.

LUPINE, BIGLEAF

Lupinus polyphyllus

TOXIC

almost 5 feet tall

blue or violet
pealike flower

distinct white
splotches on
upper flower

1 or more
unbranched
flowering stems

leaves palmately
divided 9–15
leaflets

seedpods,
hairy, 1–2
inches long

BLOOMS: June–August.

HABITAT AND RANGE: Bigleaf lupine (*L. polyphyllus*) is found in moist meadows, stream banks, and creek sides throughout the West, from southern Alaska to Quebec and south to Wyoming, Utah, Colorado, and California.

CONSERVATION: *L. polyphyllus* is common and abundant. It hybridizes quite easily and is the parent plant for many garden lupines, including the very popular Russell lupine hybrid. It frequently hybridizes in the wild with *L. perennis,* which is an essential larval plant for the endangered Karner blue butterfly. Unfortunately, the larvae will not feed on the hybrid plants, so the presence of *L. polyphyllus* is actually a threat to this butterfly population.

WILDLIFE PARTNERS: Lupines provide very important larval food for many different kinds of butterflies and bees. The leaves are bitter and slightly toxic and are generally shunned by animals.

IN THE GARDEN: *L. polyphyllus* is sometimes called garden lupine because it adapts so well to cultivation and is so easy to hybridize. Lupines make excellent garden plants and many hybrids exist. They need rich soils, full sun, and plenty of moisture.

RELATED SPECIES: Wild lupine (*Lupinus perennis*) grows 8 to 24 inches tall with light green or reddish-green stems. The Texas bluebonnet is *L. texensis.*

MEDICINAL USES: American Indians used *L. perennis* in a cold tea to stop hemorrhaging and vomiting. *Toxic.*

Archeologists believe that lupines have been cultivated for food in the Mediterranean region for over 3,000 years. Lupine beans were popular with Romans and were widely cultivated. Pliny said that nothing could be more wholesome than white lupines, eaten dry. He indicated that this would make one have fresh, glowing skin, and improve the mind and one's general outlook.

Virgil wrote of the plant and called it *Tristus Lupinus,* or the "sad lupine," presumably because if not prepared correctly, the beans are so bitter that eating them makes you sad. Most North American species of *Lupinus* are toxic. Only species native to Europe or Asia were eaten, and pickled lupine beans are still enjoyed by many Europeans.

Pickled lupine beans In the ancient theater world, actors used lupine seeds or beans as a substitute for money in plays, thus giving it one more name: *Nummus Lupinus,* meaning "money of no value."

The genus name *Lupinus* comes from the Latin *lupus,* meaning "wolf." It was thought that lupines robbed the soil of its nutrients, just as a wolf steals food. Actually, lupines, like all members of the pea family, put nitrogen back into the soil.

According to a 13th-century herbal, lupine was useful in healing the spot left after an infant's umbilical cord was cut.

Many American Indian tribes used lupine in making cloth and paper, as flavoring, and in making soap. The Menominee believed that if you rubbed this plant over your hands, it would help you control wild horses.

MINT, WILD

Mentha arvensis

4–24 inches tall

pairs of serrated fragrant leaves, 2 inches

whorls of flowers at leaf pairs, close to stem

unbranched square stem

upper lobes in 2 parts

each flower is ¼ inch pale purple, whitish or pinkish

BLOOMS: Early–midsummer.

HABITAT AND RANGE: *Mentha arvensis* prefers low-lying grassy fields and meadows, and it can spread widely in some areas. Found throughout North America, except in some southern states.

CONSERVATION: Abundant.

WILDLIFE PARTNERS: The nectar of wild mint is collected by flies, wasps, small bees, small butterflies, and skippers. The foliage, because it is so highly scented, is usually ignored by mammals.

IN THE GARDEN: Wild mint grows easily and aggressively in the garden.

RELATED SPECIES: Mountain mint (*Pycnanthemum incanum*) is common in eastern regions of North America from Ontario south to Florida. The leaves are highly scented and they are used to make tea. Medicinal usage is similar to that of wild mint (*Mentha arvensis*).

MEDICINAL USES: Wild mint tea was used by many American Indian tribes to treat fevers, stomach pain, and to quiet crying babies. The Cree mixed it with yarrow for sore or bleeding gums. The tiny mint seed is reputed as being slightly toxic for children.

Mints were named for the nymph Minthe, who was loved by Pluto. When Pluto's wife, Persephone, found out, she turned Minthe into this herb. Pluto at least gave her the unforgettable scent. Mint has also been  called *herba buena,* which means "good herb," and was dedicated to the Virgin Mary. It was customary to strew church pews with sweet-smelling plants such as mint, primrose, and violets. Brides often wore mint, woven into a garland, on their wedding day.

Close kin to European spearmint and peppermint, our native mint has glands containing essential oil that provide the strong scent. The leaves are used to flavor food and drink and to make sauces and jellies.

John Gerarde, the 17th-century botanist, said that mint mixed with salt was to be used if one was bitten by a mad dog. He also said that smelling mint would refresh the head and memory.

The Cheyenne believed mint would improve one's love life, and they used it as a part of the sacred Sun Dance ceremony.

Dried wild mint leaves make a great woodland tea or addition to bath bags or soaps.

MONKSHOOD, WESTERN

Aconitum columbianum

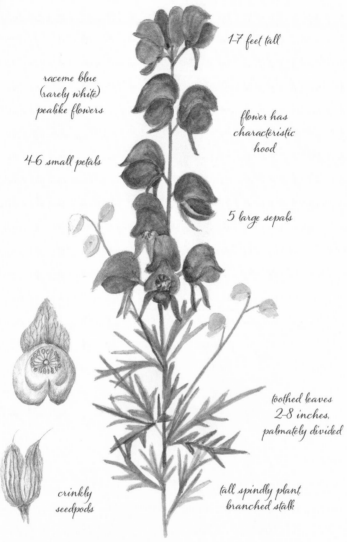

1-7 feet tall

raceme blue
(rarely white)
pealike flowers

flower has
characteristic
hood

4-6 small petals

5 large sepals

toothed leaves
2-8 inches,
palmately divided

crinkly
seedpods

tall spindly plant,
branched stalk

BLOOMS: June–August.

HABITAT AND RANGE: Grows in moist woods, hillsides, and meadows up to 9,500 feet in elevation. Found in all western states from the Pacific Ocean to the Rocky Mountains, and to South Dakota.

CONSERVATION: Though this species is secure, northern blue monkshood (*Aconitum noveboracense*) is considered threatened in its range from Iowa east to New York.

WILDLIFE PARTNERS: This plant is toxic to people and wildlife. The flowers attract bumblebees, hawk moths, and hummingbirds.

IN THE GARDEN: The plant is highly attractive and not difficult to grow in moist areas, but it is so toxic that extreme care should be exercised if you include it in a garden. It can be fatal to both humans and pets.

There are many common names for this plant, many of which include "bane," which means to kill or destroy. These include wolf's bane, leopard's bane, women's bane, blue rocket, and queen of poisons. The unusual shape of the blossom is the reason for the common name, monkshood, as well as other names such as devil's helmet. The genus name, *Aconitum,* is from the hill Aconitus, on  which Hercules was thought to have fought with Cerberus, the multiheaded watchdog of the underworld. Legend suggests that the foam from Cerberus's mouth was flung across the countryside as the monstrous dog shook his head. Where the foam landed, poisonous monkshood began to grow.

Legend says that the enchantress Medea was married to Aegeus. When she found out that Aegeus already had a son, named Theseus, she tried to kill him with monkshood put into a glass of wine.

Greeks and Romans used it on arrows and in bait left to destroy wolves. Aleut Indians used it to poison whales, placing it on spears that would paralyze them, making them easier to catch.

The European species of monkshood was thought to provide protection against werewolves.

PASSIONFLOWER

Passiflora incarnata

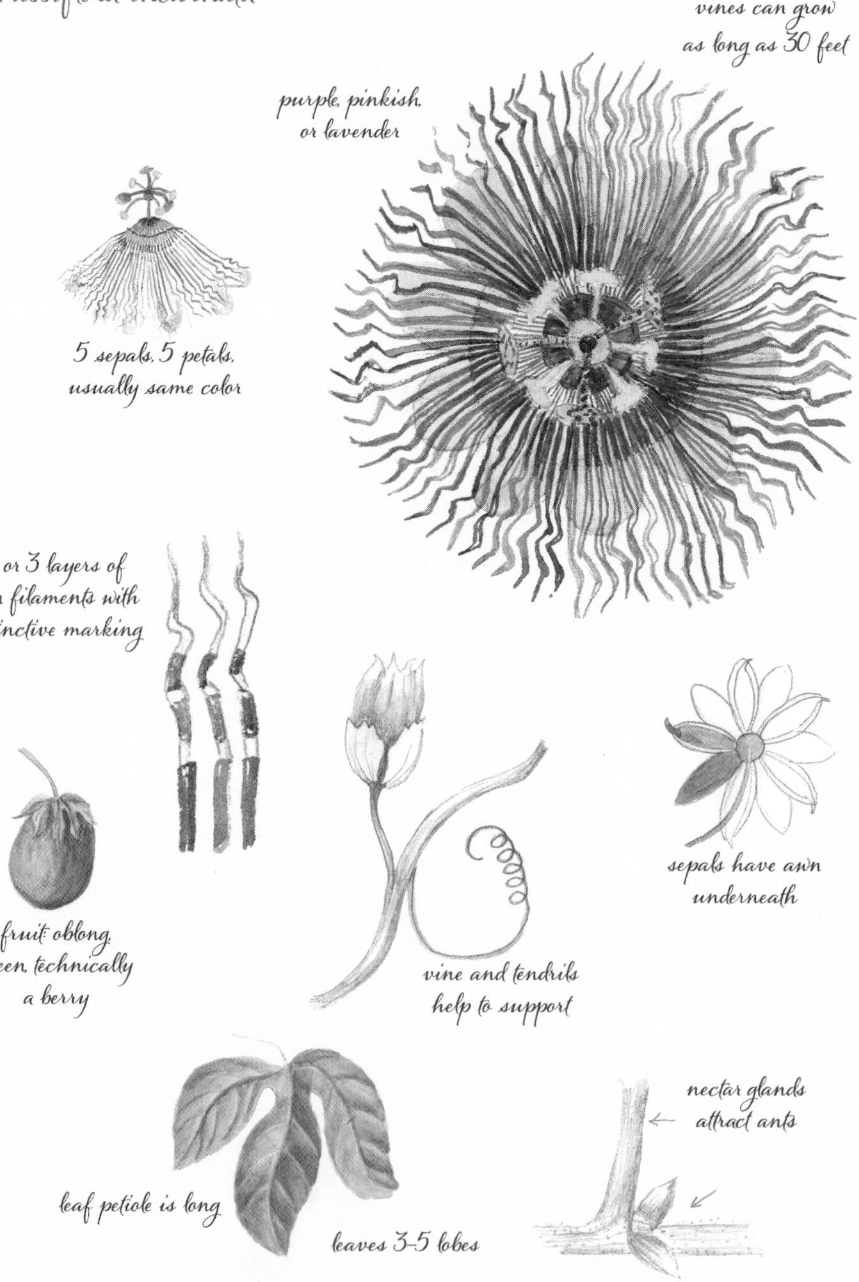

vines can grow
as long as 30 feet

purple, pinkish,
or lavender

5 sepals, 5 petals,
usually same color

2 or 3 layers of
thin filaments with
distinctive marking

fruit: oblong,
green, technically
a berry

vine and tendrils
help to support

sepals have awn
underneath

leaf petiole is long

leaves 3-5 lobes

nectar glands
attract ants

BLOOMS: July–August.

HABITAT AND RANGE: This species grows in thickets, fields, and river-banks throughout the eastern United States.

CONSERVATION: This is an aggressive grower but not a pest. Worldwide, several passionflowers are critically endangered.

WILDLIFE PARTNERS: This plant attracts a great number of pollinators, including the ruby-throated hummingbird, bumblebees and carpenter bees, and butterflies such as the crimson-patch longwing, Julia longwing, Mexican longwing, red-banded hairstreak, and the Gulf fritillary. It is the exclusive larval host for the zebra longwing.

IN THE GARDEN: This is an unusual and interesting plant to include in the garden. This perennial vine requires at least four hours of sun and rich, well-drained soil. It is great for covering a fence or trellis.

RELATED SPECIES: Other species, such as stinking passionflower (*Passiflora foetida*) and bird-wing passionflower (*P. tenuiloba*), grow in the Southwest.

MEDICINAL USE: Tea from leaves and flowers was and still is used, particularly in Europe, as a general tonic in spring, and to treat anxiety, depression, and insomnia. It was formerly approved in the US as an over-the-counter sedative, but approval was withdrawn in 1978 due to lack of testing and proof of efficacy. It was used topically by the Cherokee for burns and inflammation. The roots were used for skin irritations, wounds, and earaches.

The Cherokee name for passionflower is *ocoee*. The river and the valley in Tennessee are named this as well.

The fruit (technically a berry) is called a maypop because if you squeeze it, it makes a popping sound as it breaks open. It ripens in late summer and can be harvested when it begins to shrivel and turn soft and yellow. It is about the size of a kiwi. Local wisdom says the best way to eat the fruit is to scoop out the pulp, eat it, and spit out the seeds. An extract from the fruit and flowers was used as flavoring by both American Indians and early settlers.

SALVIA

Salvia species

There are approximately 45 *Salvia* species native to North America. Used and appreciated for everything from ornamental garden plants to an important source of food for American Indians (it was also used for sacred ceremonies), the North American salvias are both useful and beautiful.

The most well-known species is *Salvia officinalis,* an ancient herb native to Europe, which is commonly known as sage. This herb has been used in cooking and healing for many centuries. The genus name is based on the Latin *salvare,* meaning "to heal" or "to be healthy," and it refers to the healing properties of many species of the genus. A common saying during the Middle Ages was, "Why should a person die, when sage grows in his garden?"

The following includes several interesting North American species.

Chia sage, desert chia (*Salvia columbariae*)

4-20 inches tall

only 2 or 3 blooming at one time

12 or more flowers in a cluster

flowers are blue or purple

spikey calyx is bright purple

lower leaves are divided into long linear lobes, wrinkled edges

1 or 2 clusters of flowers in balls on stem

BLOOMS: March–May.

HABITAT AND RANGE: Found in chaparrals and dry slopes in California, Nevada, Utah, Arizona, and New Mexico.

This was an important food source for American Indians, as it was used to make pinole. (Pinole, which is made from corn, cocoa, agave, cinnamon, chia seeds, and spices, was used as a nutritious base for beverages, breads, or cereals.) This plant is *not* the main source for commercially available chia seeds, which come from a South American species, chia (*Salvia hispanica*).

White sage, bee sage, sacred sage (*Salvia apiana*)

DESCRIPTION: Attractive whorls of evergreen gray and green leaves with small white or light purple flowers on a small shrub, up to 5 feet tall.

BLOOMS: Spring.

HABITAT AND RANGE: Found on dry slopes and chaparral of Southern California.

CONSERVATION: The overharvest of white sage is of great concern, particularly to American Indians. Though not endangered, the health of the population depends on controlling illegal harvest.

Historically, white sage was used in sacred ceremonies by indigenous peoples throughout the Southwest and was burned as incense for purification. Fires made from the foliage of white sage were made to assure good hunting. Branches were burned in the homes of those who were sick to help purify the air. The Luiseño of Southern California peeled the new stems and ate them raw. The Cahuilla used sage tea for fevers and coughs. They also used it as a shampoo and an underarm deodorant to mask the human smell before a hunt.

Mealy blue sage, mealy cup sage (*Salvia farinacea*)

DESCRIPTION: Spikes of densely packed bright or light blue flowers make this a great garden plant. It is easy to grow and requires little water—too much moisture results in sprawling, weak stems. The leaves are bright and shiny. Grows 1 to 3 feet tall. The sepals are covered with thick hairs, giving this plant its common name because it looks as if the plant is covered with a dusting of flour or meal. This is a great plant for attracting hummingbirds and bees.

BLOOMS: Summer

HABITAT AND RANGE: Native to Texas and Oklahoma in fields and open spaces.

CONSERVATION: Abundant

Lyreleaf sage (*Salvia lyrata*)

DESCRIPTION: The basal leaves of this plant are large and variable, many being lobed with some resemblance to a lyre. It is a short-lived perennial and the flower stalk, with bluish-purple flowers, grows to a height of 1 to 3 feet.

BLOOMS: Spring–early summer

HABITAT AND RANGE: Common in fields and waste places throughout its range from Connecticut south to Florida, west to Texas, and north to Missouri.

CONSERVATION: Abundant

Mourning doves are reported as eating the seeds, and the nectar is good for many kinds of bees. The root was used by the Catawba to treat sores and wounds, and the leaves and flowers were made into a tea for treating coughs and colds by the Cherokee. Folk healers at one time hoped that this plant would be an effective cancer cure, but it did not prove to be the case.

Red sage, Texas sage (*Salvia coccinea*)

DESCRIPTION: This plant features loose whorls of flowers on a square stem. Individual flowers are about 1 inch long and have two stamens. It grows 1 to 3 feet tall.

BLOOMS: Spring–fall.

HABITAT AND RANGE: This plant grows in open, sunny areas in eastern and southern Texas.

CONSERVATION: Abundant.

This is a popular gardening plant as it is easy to grow, offers bright red flowers, and attracts bees and especially hummingbirds. Other common names are scarlet sage, blood sage, Indian fire, and hummingbird sage.

VIOLETS

V. canadensis

V. palustris

V. pedate

V. sororia

V. pubescens

GENERAL DESCRIPTION: There are 87 to 90 species of violets native to the United States. Most are bilaterally symmetrical with a lower petal, a hollow spur, and a wide landing pad for insects. Leaves are most often scalloped and heart-shaped. Common colors include blue and violet, yellow and white, with an occasional bicolor or pink. It's relatively easy to determine whether or not it's a violet, though it's much more difficult to identify to the species level as they interbreed in the wild quite readily.

HABITAT AND RANGE: Violets grow in several kinds of areas, but generally, they prefer shady woods. Violas are native to every state, including Hawaii.

CONSERVATION: There are several *Viola* species considered endangered or threatened. Johnny-jump-up (*V. pedunculata*), which is native to western North America, is the primary larval host plant for the endangered Callippe silverspot butterfly. On the other hand, common violets are often considered invasive.

WILDLIFE PARTNERS: Violas are the larval host plant for greater fritillary butterflies. Seeds are eaten by mourning doves, wild turkeys, and bobwhite quails.

IN THE GARDEN: Small but lovely, many *Viola* species make great garden plants. Common varieties can become invasive.

Violets are the state flower for Illinois, New Jersey, Rhode Island, and Wisconsin.

In mythology, the nymph Io was loved by Zeus. To hide her from his jealous wife, Hera, Zeus changed Io into a white cow. Not accustomed to eating rough grass, Io began to cry. Zeus changed her tears into the violet.

Women of ancient Rome mixed violets in goat's milk and used it on their skin. Pliny recommended violets to induce sleep, strengthen heart muscles, and calm anger. Wearing a garland of violets was thought to (hoped to) prevent drunkenness.

During the exile of their leader, the French Bonapartists chose the violet as their symbol. Napoleon promised to return when the violets bloomed in spring, and on March 20, 1815, he returned to Paris.

Shakespeare used it as a symbol of modesty and constancy in love, and he included it in many sonnets.

Violet leaves are high in vitamins A and C and can be eaten raw or cooked. The leaves become mucilaginous when cooked and were used for thickening sauces and stews by both Europeans and early American settlers. The flowers are quite edible, and they are sometimes crystallized to make candy. Folk healers believe that violet tea helps get rid of a headache.

GARDENING WITH NATIVE PLANTS

Although it's thrilling to find beautiful wildflowers growing in the wild, it's also fun and satisfying to grow them in your own yard or garden. Growing native plants offers many benefits. They:

- provide native plant material for pollinators and food for larvae and other wildlife
- require fewer resources and maintenance if grown in places that mimic a natural environment
- help to maintain and preserve important natural ecosystems

Many gardeners are accustomed to choosing plants based on USDA horticultural growing zones, which are determined by minimum temperatures generally experienced in a particular area. When choosing native plants, though, you might consider choosing plants based on natural ecosystems instead. (See the map on page 10.)

As a general rule for best results in growing native plants, mimic a plant's natural environment as closely as possible.

WHAT DOES "NATIVE" REALLY MEAN?

If you want to be a purist and grow *only* indigenous plants, you'll have to determine how pure you want to be. Just because a plant is part of a general ecosystem, it does not mean that it is necessarily native to *your* particular part of that region. You'll have the best luck and provide the greatest use for pollinators by planting plants that are indigenous to regions close to where you live and are acclimated to the specific conditions of your garden.

Many native plants have now been hybridized to offer gardeners a wider variety of color and size of bloom. But once a particular species has been genetically altered, is it really still a native plant? For some people, it's important to only grow plants that are as genetically close as possible to those that grew there before the arrival of the Europeans in North America. Others will welcome the new, perhaps more colorful and bigger, varieties now available.

Planning a native plant garden is fun. Do a little research. Take a notebook and go visit other gardens, hike in wild and scenic areas close to where you live, and take notes of the most beautiful plants you see. Do remember that the most prolific plants may not be native. Invasive, exotic plants have a way of forcing out many natives, so be sure of what you're looking at. Only native species are included in this book. You may not be able to grow all the things you want to grow, and it may take years to get a profusion of your favorites, but with patience and a little effort, you will be rewarded with a slice of the wild right outside your doorstep.

There are countless ways to include native plants in your landscape. The easiest may be to simply plant a few in and amongst flowers that you already have established. Or you may want to create a new garden or space and include just the natives. There is no right or wrong way; it just depends on your own personal dreams and desires.

WHERE TO GET PLANTS

Don't dig from the wild. It may be tempting, but getting wildflowers from the wild is generally impractical since so few of them, especially woodland species, transplant successfully. But, of course, the most important reason not to dig from the wild is the ethical one. We are losing populations of wild plants at an alarming rate, primarily due to loss of habitat and the threat of invasive species. We should all be working hard to preserve each and every native plant growing in the wild.

Fortunately, there are many commercial nurseries in every region of the country that specialize in or at least carry native plants. It's best to "buy local." A nursery close to you will be more likely to carry plants native to your region, and these plants will be already adapted to your climatic conditions.

When you purchase plants, make absolutely certain that they were propagated and not dug from the wild. It's relatively easy to tell by taking the plant out of the pot. If it looks like potting soil, you can be assured that the plant was propagated. If it looks uneven, like native soil, and the roots look cut off, you probably don't want to buy it. If you are in doubt, ask the nursery.

Links for finding native plant nurseries are found in the Appendix of this book.

EVALUATING YOUR GARDENING SITE

Before you begin putting in plants, it's important to know what conditions you have.

SOIL: You can find plants that grow in just about any soil, but they may not necessarily be the ones you *want* to grow. Soil composition can be altered with amendments. If you want to grow woodland plants, you'll probably want to add organic material to your soil. If plants require well-drained soil, you'll need to dig deep and add materials to help with drainage. Remember that compost is always a terrific soil amendment.

The pH and composition of soil should also be considered. If you're not sure about your soil, have it tested. Almost all states offer this service for a nominal fee. Along with the results, you will receive suggestions for adding nutrients to the soil or for lowering or raising the pH level. It is a very useful test and may save you a lot of frustration down the road.

Remember that your soil is an entire ecosystem all to itself, rich with organisms. Most of these organisms are too small to see, but they are an integral part of healthy soil. An old gardening adage is to "feed the soil, not the plant," and it's an important phrase to keep in mind when gardening with natives as with cultivated plants, perhaps more so. Be careful what you "feed" your soil. Just like providing good nutrition for your children, you want to give your soil elements that are high in value and low in chemicals. Chemical fertilizer may temporarily boost growth of a particular plant but may eventually destroy all those wonderful, beneficial organisms in the soil and do lasting damage. Fortunately, organic fertilizers are readily available.

SUN AND SHADE: How much sunlight does your site receive? Plants that require "full sun" will not be satisfactory if they don't receive enough light. They may grow, but generally sun-loving plants need a lot of sun to bloom well. Conversely, plants that do well in shady areas will dry out or wilt with too much sun.

Take note of when and where the light comes from. Deciduous woodland light is seasonal, and it occurs before leaves come out in spring, which is when most forest wildflowers bloom. If you're planting under evergreens, your light situation will be different.

WATER: Determine the closest water source to your site. Different plants require different amounts of moisture, but almost all new plantings will need irrigating until the roots are established. Know the water requirements for the plants you're going to grow. In general, field flowers require a lot less water than woodland flowers.

FEASIBILITY: Once you've answered some questions about your site, determine whether or not it's really feasible to use. How much will you have to work with the soil to provide the right conditions for the plants you want to grow? How much sun does it receive? How far is it from a watering source? How accessible is it? Can you easily get plants, tools, and materials to it? Gardening is a lot of work, so be sure to put your efforts into a place that you can see and enjoy.

REGIONAL FAVORITES

Plant communities don't adhere to strict boundaries. There is a lot of overlapping and mixing and mingling. But wherever you live, you can plant flowers that are native to your region. The following suggestions are loosely based on the following geographical regions: Midwest, Northeast, Pacific Northwest, Southeast, Southwest, and Rocky Mountains.

MIDWEST

SUN: Purple coneflower (*Echinacea purpurea*), sunflower (*Helianthus annuus*), black-eyed Susans (*Rudbeckia hirta*), goldenrod (*Solidago* sp.), blue wild indigo (*Baptisia australis*), prairie rose (*Rosa* sp.), penstemon (*Penstemon* sp.), prairie smoke (*Geum triflorum*), blazing star (*Liatris spicata*).

SHADE: Cardinal flower (*Lobelia cardinalis*), Solomon's seal (*Polygonatum biflorum*), black cohosh (*Actaea racemosa*), columbine (*Aquilegia canadensis*), Jack-in-the-pulpit (*Arisaema triphyllum*), wood lily (*Lilium philadelphicum*).

NORTHEAST

SUN: New England aster (*Symphyotrichum novae-angliae*), coreopsis (*Coreopsis tinctoria*), purple coneflower (*Echinacea purpurea*), milkweed

(*Aesclepias* sp.), wild geranium (*Geranium maculatum*), phlox (*Phlox paniculata*), goldenrod (*Solidago* sp.).

SHADE: Bellwort (*Uvularia grandiflora*), bloodroot (*Sanguinaria canadensis*), eastern columbine (*Aquilegia canadensis*), dwarf crested iris (*Iris cristata*), mayapple (*Podophyllum peltatum*), Jack-in-the-pulpit (*Arisaema triphyllum*), trout lily (*Erythronium americanum*), Solomon's seal (*Polygonatum biflorum*).

PACIFIC NORTHWEST

SUN: Skyrocket (*Ipomopsis aggregata*), heartleaf arnica (*Arnica cordifolia*), arrowleaf balsamroot (*Balsamorhiza sagittata*), showy fleabane (*Erigeron speciosus*), wild rose (*Rosa* sp.), goldenrod (*Solidago* sp.), Cascade penstemon (*Penstemon serrulatus* or *Penstemon fruticosus*), yarrow (*Achillea millefolium*), camas (*Camassia quamash*), Douglas aster (*Aster subspicatus*), Douglas iris (*Iris douglasiana*).

SHADE: Shooting star (*Dodecatheon* sp.), Oregon fawn lily (*Erythronium oregonum*) bunchberry (*Cornus canadensis*), western columbine (*Aquilegia formosa*), trillium (*Trillium* sp.), violet (*Viola* sp.).

SOUTHEAST

SUN: Phlox (*Phlox paniculata*), black-eyed Susans (*Rudbeckia hirta*), purple coneflower (*Echinacea purpurea*), bee balm (*Monarda didyma*), butterfly-weed (*Aesclepias tuberosa* and other *Aesclepias* species), goldenrod (*Solidago* sp.), New England aster (*Symphyotrichum novae-angliae*), joe-pye weed (*Eutrochium maculatum*), evening primrose (*Oenothera biennis*).

SHADE: Woodland phlox (*Phlox divaricata*), bloodroot (*Sanguinaria canadensis*), eastern columbine (*Aquilegia canadensis*), dwarf crested iris (*Iris cristata*), mayapple (*Podophyllum peltatum*), Solomon's seal (*Polygonatum biflorum*), bellwort (*Uvularia grandiflora*), violet (*Viola* sp.), Jack-in-the-pulpit (*Arisaema triphyllum*), Turk's cap lily (*Lilium superbum*).

SOUTHWEST

SUN: California poppy (*Eschscholzia californica*), wild lupine (*Lupinus perennis*), prickly pear cactus (*Opuntia polyacantha*), banana yucca (*Yucca* baccata), plains coreopsis (*Coreopsis tinctoria*), California goldenrod (*Solidago californica*), evening primrose (*Oenothera biennis*), firecracker penstemon (*Penstemon eatonii*).

SHADE: Yellow columbine (*Aquilegia chrysantha*), scarlet monkey flower (*Erythranthe cardinalis*), bellflowers (*Campanula rotundifolia*), skyrocket (*Ipomopsis aggregata*), New Mexico yellow flax (*Linum neomexicanum*).

ROCKY MOUNTAINS

SUN: Fireweed (*Chamaenerion angustifolium*), Lewis's monkey flower (*Erythranthe lewisii*), wild rose (*Rosa* sp.), Nuttall's larkspur (*Delphinium nuttallianum*), wild blue flax (*Linum lewisii*), camas (*Camassia quamash*), bear grass (*Xerophyllum tenax*), wild lupine (*Lupinus perennis*), showy milkweed (*Asclepias speciosa*), coreopsis (*Coreopsis tinctoria* or *Coreopsis lanceolata*), Colorado columbine (*Aquilegia caerulea*), yarrow (*Achillea millefolium*).

SHADE: Rocky Mountain Iris (*Iris missouriensis*), Lewis's monkey flower (*Erythranthe lewisii*), shooting star (*Dodecatheon sp.*), Rocky Mountain penstemon (*Penstemon strictus*), smooth aster (*Symphyotrichum laeve*), windflower (*Anemone multifida*).

FORAGING AND COOKING WITH WILDFLOWERS

For the most part, I have included only native plants in this book because I think we need to understand and celebrate the plants that are uniquely ours. But in this chapter about picking and harvesting, I include a few common weeds. These plants often outcompete the natives for space and resources, so eliminating them from the ecosystem is actually beneficial.

There are three important rules about foraging for edible plants.

The first and most important rule: Do no harm to yourself. Know, without question, the identity of the plant and when to harvest it. If you have any question *at all* about whether or not to eat a plant, *don't!* It's just not worth it.

The second rule: Even if you are certain of the edibility and identification of a plant, if you've never eaten it before, start small and consume only tiny amounts at first. Each body is different and will react to new foods in different ways. And if it tastes bad to you, don't eat it.

The third rule: Do no harm to the environment. Don't harvest anything unless it's plentiful and abundant, and harvest sparingly the parts of a plant that do it the most damage. In other words, feel free to pick some leaves off plants, but don't dig many up by the roots.

This list includes easily identifiable weeds and wildflowers that are safe to eat. Most of these plants are native to or naturalized in almost every region of the United States. This is by no means a comprehensive list, but just a beginner's guide. It will, hopefully, pique your interest in foraging our native plants and in learning more about the harvest waiting for you in our fields and woodlands.

Before you eat anything, be sure you know what you're picking, clean it carefully, eat only the parts recommended, and cook it when recommended. The recipes that follow are easy and fun. Bon appétit!

PLANT	PARTS USED	HOW TO USE
Bee balm (*Monarda didyma*)	Leaves	Leaves cooked or in tea
	Flowers	Garnish, in jelly, flavoring
Cattail (*Typha latifolia*)	Young stalks	Eaten raw or sautéed
	Roots	Peeled, dried, ground as flour
	Pollen	Mixed with flour and meal
Cactus (*Opuntia polyacantha*)	Fruit	Spines removed, raw or cooked, made into jelly
Chickweed (*Stellaria media*)	All plant parts	Raw or cooked like potherb
Chicory (*Cichorium intybus*)	Young leaves	Raw
	Flowers	Garnish
	Root	Roasted and ground into coffee substitute
Dandelion (*Taraxacum officinale*)	Leaves	Raw or cooked like potherb
	Flowers	Battered, fried like a fritter, flavor for wine
	Unopened buds	Pickled like capers
Fireweed (*Chamaenerion angustifolium*)	Young shoots	Raw or sautéed
	Young leaves	Raw or cooked like potherb
	Young flowers	Garnish, made into jelly
Lamb's-quarter (*Chenopodium album*)	Young leaves	Raw, cooked like potherb
	Young shoots	Raw or sautéed
	Seeds	Roasted, in breads or soups

PLANT	PARTS USED	HOW TO USE
Mint (*Mentha arvensis*)	Leaves	Raw or dried for tea
	Flowers	Raw, garnish, jelly, or tea
Onion (*Allium cernuum*)	Leaves	Garnish or flavoring
	Bulbs	Peeled, cooked for flavoring
Passionflower (*Passiflora incarnata*)	Fruit	Peeled, seeded, raw, in jam, or to flavor drinks
Plantain (*Plantago* sp.)	Leaves	Sautéed, cooked like potherb
	Seed heads	Sautéed or added to muffin or bread dough
Rose (*Rosa* sp.)	Blossoms	Raw in garnish, cooked in syrup, flavoring, candied
	Hips	Dried: jam, jelly, flavoring for tea
Strawberry (*Fragaria virginiana*)	Leaves	Dried, made into tea
	Fruit	Raw, made into jams and sauces
Sunflower (*Helianthus annuus*)	Seeds	Raw, roasted, extracted for oil
Violet (*Viola* sp.)	Leaves	Cooked like potherb, thickener to soups and stew
	Flowers	Raw as garnish, candied, made into jelly
Yucca (*Yucca filamentosa*)	Buds and flowers	Raw in salads, fried like a fritter

For an illustration and more information and about the *native* plants in this list, refer to their entries within the book. The non-native plants are described and illustrated below.

CHICKWEED (*STELLARIA MEDIA*): A sprawling, weak-stemmed weed with small oval, opposite leaves, and white star-shaped flowers that appear at the end of the stems. Grows 6 to 15 inches tall.

CHICORY (*CICHORIUM INTYBUS*): Basal rosette of leaves, 3 to 10 inches long, with hairy red mid-rib. Flower stalks up to 4 feet tall with sky blue blossoms that open in the morning and close by midday.

DANDELION (*TARAXACUM OFFICINALE*): Basal rosette of deeply lobed leaves and a hairless mid-rib. Bright yellow flowers at the end of a hollow, smooth stalk. Seed heads are composed of fluffy seeds attached to long silky strands. Although there are several plants that resemble dandelions, none of these are poisonous. Harvest leaves before flowering.

LAMB'S-QUARTER (*CHENOPODIUM ALBUM*): At full height, this plant is 3 to 5 feet tall with many branches. Leaves are diamond-shaped and velvety, whitish underneath. Green flowers turn reddish-brown in fall and grow on short stalks at the ends of the branches. Harvest basal leaves when they are very small early in spring.

PLANTAIN (*PLANTAGO SP.*): Long, narrow leaves grow in a basal rosette. Seed heads are small, elongated, and brown on leafless stalks. Harvest leaves very early in spring. Seed heads are best when greenish-brown.

RECIPES

The following recipes can be adapted for use with many different plants, depending on what's available and abundant.

Herbal tea

Place 4 tablespoons fresh, or one tablespoon dried, plant material in a tea strainer. Pour one cup of boiling water over this and let steep for 4 to 15 minutes, depending on your personal preference. Strain and sweeten with honey, if desired.

Note: Try the leaves and/or flowers of bee balm, mint, rose, violets.

Floral syrup

1 cup water
1 cup edible flower petals (or wild berries)
1 cup sugar
Splash of vanilla or sprinkle of cinnamon, optional

Mix the water with the petals (or berries). Heat for 20 minutes on medium heat, then strain. Use 1 cup of this flavored water with 1 cup sugar and cook over low heat until sugar has dissolved. Add a splash of vanilla or a sprinkle of cinnamon if desired. Use over ice cream, cakes, or pancakes.

Note: Try mint, chicory flowers, rose petals, violet flowers, or bee balm petals.

Salad

Mix greens of dandelion, lamb's-quarter, chickweed, and plantain, topped with pieces of cattail, pickled dandelion buds, and chopped wild onion. Garnish with chicory and violet blossoms. Dress with oil and vinegar for an unforgettable spring treat.

Rose hip tea

Pick rose hips when fully ripe in fall. Cut into pieces and dry. When thoroughly dry, store in an airtight container. For tea, take a tablespoon of the dried hips, steep in 1¼ cups boiling water for 15 minutes. Sweeten with honey.

Note: This tea is also nice blended with other herbal teas.

Pickled dandelion buds

½ salt

1 quart water

1 cup unopened dandelion flower buds

Spicy vinegar (white vinegar infused with bay leaves, garlic cloves, and peppercorns)

Make a salt water solution: In a bowl, combine the salt with the water.

Pack the buds into a glass jar and pour the salt water solution over it. Leave for 24 to 48 hours, drain and cover with clean, unsalted water for another 24 hours. Drain, dry, and place the buds in small jars. Heat the spicy vinegar and pour it over the buds. Put clean lids on the jars and process in a hot water bath (as you would for jams and jellies) for 15 minutes.

Crystallized violets

Pick perfect flowers with short stems still attached. Clean and dry thoroughly. Make a meringue by beating egg whites until almost dry (if you plan to eat these, use a meringue powder available at specialty food stores to avoid ingesting raw eggs). Dip the blossoms in the meringue, then use a sifter to sprinkle superfine granulated sugar over the blossoms. Place on a cookie sheet and put in a warm, dry place and allow to harden. Use as a garnish on cakes and cupcakes.

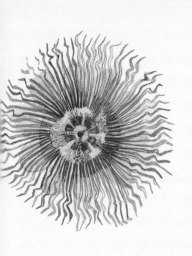

CRAFTING WITH WILD PLANTS

The following activities or crafts are fun and generally pretty simple. Many of them are more decorative than utilitarian, but they will hopefully offer you the joy of working with your hands and a new appreciation for the wild plants that surround you. As you do these activities, you will also, undoubtedly, be more grateful for all our modern conveniences such as electric stoves, pots and pans, and sewing machines.

RULES FOR CRAFTING

The same essential rules apply whether you're picking plants or flowers to eat or to craft with.

1. Don't harm yourself.
2. Don't harm the environment.

All the wildflowers (and weeds) suggested for these activities are common, harmless, and abundant enough to fulfill rule #2. Just be certain of their identity to fulfill rule #1. Have fun!

USEFUL FLOWERS AND WEEDS

Aster: pressed, pounded (see page 275), cut flower

Bee balm: potpourris, soaps, infusions

Blue wild indigo pods: decorative pods for arrangements

Cattail: weaving

Coneflower, purple: flowers and roots for skin ointments

Coreopsis: dye, pressed, pounded, cut flower

Fireweed: pressed, pounded

Forget-me-not: pressed

Goldenrod: dye, cut flower, dried

Larkspur: pressed, cut flower, pounded

Milkweed pods and seeds: stuffing, decorative

Passionflower: soaps, oils, infusions

Phlox: cut flower, pressed, bouquets

Rose: cut flower, pressed, bouquets, salves, soaps

Sunflower: seeds, arrangements, dye, facial creams

Violets: pounded, salves, soaps

Yarrow: soaps, infusions, bath oils, arrangements

PROJECTS
SKIN AND BATH PRODUCTS

The best plants to use for these projects: roses, violets, yarrow, echinacea, mint, passionflowers.

Floral infusion
Distilled water (1 cup for each ¼–½ cup plant material)
Petals or leaves, fresh, clean (¼–½ cup for each 1 cup distilled water)

In a nonreactive pot, bring the distilled water to an almost-boil. Add the petals or leaves. Turn off the heat and steep for 30 minutes. Allow to cool and store in glass jars.

Bath oil
2 ounces almond or sesame or walnut oil
12 drops of essential oil (lavender or rose)
4 tablespoons floral infusion

Mix all the above ingredients together. Store in a glass jar. Add 1 tablespoon per bath.

Bath bags
¼ cup oatmeal
¼ cup dried leaves and petals of rose, violet, coneflower, fireweed, passionflower
   blossoms, sunflower petals (or ½ cup fresh plant material)
4 drops essential oil (lavender or rose)
Cheesecloth or oversized muslin tea bag

Toast ¼ cup oatmeal. Allow to cool and mix with the dried petals and leaves and essential oil. Place in a large square of cheesecloth. Tie securely and place in the bath.

Floral infusion soap

8-ounce bar clear glycerine soap, grated
4–6 drops essential oil (rose, lavender, rosemary)
¼ cup floral infusion
¼ cup toasted oatmeal
Almond or sesame oil for greasing molds

Microwave the grated glycerine on medium for 45 seconds or until melted. Heat the floral infusion until quite warm, then pour this into the melted soap and stir until smooth. Add the essential oil and oatmeal and mix again. Pour into greased molds (though fancy soap molds are available, you can just put this into an ice tray or anything else!). Allow to cool and harden.

Violet hand cream

1 tablespoon beeswax
6 tablespoons almond oil
2 tablespoons violet floral infusion

Place the beeswax and oil in a medium bowl. Place in the microwave and heat until warm (start with 30 seconds and add time as needed). Add the violet infusion and whisk until creamy. Cool. Pour into small clean bottles and enjoy—or share!

DYEING AND WEAVING PROJECTS

Dyeing wool yarn

Flower suggestions: Coreopsis, asters, sunflowers, goldenrod

PREPARING THE DYE BATH: Use equal weights of flower and yarn. For example, an 8-ounce skein of wool yarn will need 8 ounces (1 cup packed) flowers or petals. Cover the flowers with water and bring to a boil. Turn off the heat and allow to sit 2 hours. Strain out the plant material, leaving the dyed liquid. Wet the yarn with plain water.

MAKE THE MORDANT: Use 4 tablespoons alum and 4 tablespoons cream of tartar. Place in 1 gallon water. Heat but don't boil. Add the wet yarn. Saturate the wool and leave 15 minutes.

DYEING: Take the yarn out of the mordant bath and put in the dye bath. Heat but do not boil, then turn off the heat and allow to sit between 2 to 24 hours, depending on the intensity desired. Rinse with cool water until the water is clear.

Weaving

Several plants can be used for weaving into baskets. Cattail leaves are probably the most readily available and easiest to use. If you're new to weaving, you might think about weaving a flat mat instead of trying to shape it into a basket. Whichever you do, make sure that the leaves stay damp and pliable. If they dry out, they tend to break.

MATERIALS: An armful of green cattail leaves, cleaned and thoroughly dried. When you're ready to use them, submerge them in water or wrap them in a damp towel for several hours to make them pliable.

HOW TO DO IT: Begin with four leaves in an over-and-under pattern in the center of the leaves. Add another leaf to one side, turn the mat and add it to the next (adjacent) side. Continue, working around the mat with the same over-and-under pattern. Be sure to keep it square as you weave the leaves as tightly as possible. Keep the leaves damp and pliable—you may have to wrap the mat in a wet towel again. When you have about 5 inches left, begin weaving the "tails" into the back side to create an edge. Do this to all the leaf ends until your mat is complete. Store in a cool, dry place.

USING FRESH WILDFLOWERS

Picking flowers from the wild is a glorious and wonderful thing to do as long as you're picking flowers that are common and abundant, and you aren't robbing pollinators of flowers that they can use later in the season. Pounding wildflowers is the art of directly transferring a flower image to paper or cloth. For instructions, see page 275. These treasures from the fields and woodland borders can be used for a variety of fun projects and crafts. Many of these are simply suggestions rather than elaborate crafts. Have fun!

Cake decorations

Pick a few beautiful edible flowers (for a list, see Foraging and Cooking with Wildflowers, page 256).

MATERIALS: Wildflowers, cake, straws.

HOW TO DO IT: Cut the stems off the wildflowers, so they are about 2 to 3 inches long. Bake (or buy) a plain iced cake. Put the straws into the center top of the cake. Slip the flower stems into the straws. The flowers are not in water and won't last a long time, but they make a stunning presentation while they do last!

Bouquets

For a wedding, a party, a dress-up game, or just for fun!

MATERIALS: Wildflowers, ferns, leaves, ribbons, flower bouquet holder (essentially a funnel with lace around it, available at craft stores), and white floral tape.

HOW TO DO IT: Use ribbons that are the same colors or complementary shades to the flowers. Make a large bow and either glue or sew it to the lace surrounding the funnel of the holder. Cut the wildflower stems to about 2 to 3 inches long. Place them flat on a table and place ferns or leaves around them until you are satisfied with the arrangement. Wrap all the stems together, using floral tape. Place in water. When you are ready to carry the bouquet, place the flower stems through the bouquet holder. Rewrap with white floral tape if necessary.

Note: A floral headpiece can be done in a similar way. Use ribbon to cover a wire ring that will go around your head. Wrap small bundles of blossoms together and attach these to the headpiece with floral wire. When you're done, be sure to cover all the wire with white floral tape. Weave a narrow ribbon through the flowers and allow it to trail down the back.

DRIED FLOWER CRAFTS

You can use dried wildflowers either whole in arrangements or decorations, or you can just use dried petals and leaves. No matter what your end product is, you want to make sure that your plants are thoroughly dried before you begin crafting with them. If they retain moisture, they might mold. Please refer to the chapter Wildflowers and Children (page 272) for quick and easy steps for drying flowers, petals, and leaves.

Pressed flowers

Many common wildflowers press well and hold their color. For whole blossoms, try violets, aster, coreopsis, goldenrod, phlox, ferns, flax, wild geranium, larkspur, or roses. For individual petals, try arnica or balsamroot, sunflower, black-eyed Susans, or just about anything else that is relatively thin, flat, and abundant.

There are commercially available plant presses, but if you don't have one available, just use blotting paper (or paper towels) between the pages of a large, heavy book. Take small sticky notes to mark the pages and identify the plants.

HOW TO DO IT: Pick the flowers in their prime and place them carefully in the prepared book or press. Be sure that they are lying as flat as possible and are in a graceful position. Leave for several days before you check on them.

Once dry, you can use them to decorate all kinds of things. Create a flat "arrangement" and cover with glass and frame. Use individual blossoms on place cards, covered with clear packing tape for a dinner party. Cover an entire sheet of heavy paper with an artful display of flowers and cover with clear contact paper. If it turns out well, photocopy this sheet to use as wrapping paper, note cards, or stationery.

MAKE YOUR OWN ILLUSTRATED WILDFLOWER BOOK (WITHOUT DRAWING A THING!): Collect and press a variety of (common and abundant) flowers that you see, then place them on sheets of heavy paper. Carefully glue them down (or cover with clear contact paper). Write notes in the margins about where and when you saw it, and add other details too (a poem? notes about the occasion? personal comments?). Place these pages into a folder or book for a beautiful reminder of your experiences as a naturalist!

WILDFLOWERS
AND CHILDREN

Crafts are a really fun way to help kids fall in love with wildflowers. Here are suggestions for a few fun and very easy projects to do with children.

Be sure you know the identity of plants you're using. Don't pick anything that is potentially harmful or that is scarce or rare. And don't pick plants that have been sprayed with chemicals. For these crafts, you can also use garden plants or go for the weeds! You'll actually be helping the environment by picking them.

USING FRESH FLOWERS, PETALS, AND LEAVES

Flower candy bark

MATERIALS: White (or milk) chocolate chips. Edible (and colorful) flower petals (beebalm, violets, dandelions blossoms, rose petals). Parchment paper. Rimmed baking dish.

HOW TO DO IT: Place 1 cup chocolate chips in a microwave-proof bowl. Melt in the microwave at 50 percent power for 30 seconds. Take out and stir. Continue melting at 30-second intervals until all the chips are melted. Line the rimmed baking dish with parchment, pour the melted chocolate, and spread into a thin layer. Immediately sprinkle with flower petals. Place in the freezer or refrigerator until hard and break into pieces. Lovely! And yummy!

Printmaking

You'll be using flower petals and leaves as crayons.

MATERIALS: Regular white paper. Colorful petals. Green leaves.

HOW TO DO IT: Even very young children can do this. Simply take a piece of white paper and rub the plant material directly on the paper. Make shapes and designs or just "scribble."

Stamping

MATERIALS: Thick (non-shiny) white paper. Leaves or petals with prominently raised veins, such as ferns. Acrylic paint. Paintbrush or small foam roller. A pie pan or paper plate. Paper towels.

HOW TO DO IT: Squeeze the paint onto the plate, add a little water, and mix until smooth. Using the foam roller, put paint evenly on the entire surface of your plant piece. Use the side with the most raised details. Carefully, place the plant piece, paint side down, on a piece of thick paper. Cover with a paper towel and press straight down. Don't rub back and forth. When you have pressed it completely, remove the paper towel, and carefully pick up the plant. Voilà!

Pounding flowers

MATERIALS: Cotton or silk fabric or absorbent paper (such as used for watercolor or printmaking). Hammer. Wooden chopping board. Paper towels Colorful, small flowers that are relatively flat (like a violet) or leaves that have an interesting shape (ferns are great!).

HOW TO DO IT: Place a paper towel on the chopping board. Put the fabric on top of this, then add your plant material. (Start small!) Try a piece of fern. Place another paper towel on top of the plant. Hold it so it will stay as stationary as possible, but watch your fingers! An adult will probably want to hold it to allow a younger child to do the hammering. Hammer so that you hit all the plant material underneath the paper towel. When you think you've thoroughly hammered the plant, uncover and see the magic.

Note: Some plants hammer amazingly well, transferring details. Others, especially if they are thick or multilayered, just squish. If you need to, take apart the plant and hammer them in a single layer. Try violas, pansies, vinca, ferns, rose petals.

Using plants to make (temporary) figures and designs

MATERIALS: All kinds of petals, grass, leaves, ferns, flowers, buds. White paper.

HOW TO DO IT: This is a temporary but fun activity—and you can always photograph your creation to make it last. Simply create designs or figures with the plant material. If you are doing a figure, think about making a torso, a skirt, arms, legs, feet, a face, and hair. You'll be amazed at how beautiful they can be. Have fun.

USING DRIED FLOWERS, PETALS, AND LEAVES

Place plant material on a rimmed baking dish in a warm 200°F oven for 30 to 40 minutes. Plants have to be completely dry before use, otherwise they'll begin to mold.

TRY: Roses, bee balm, violets, purple coneflower, black-eyed Susans, sunflower petals, goldenrods, strawberry leaves and blossoms, yarrow, fireweed, geranium, phlox, asters, and passionflower.

Potpourri

MATERIALS: Essential oils (lavender, rose, or violet). Small cloth bag or cheesecloth. String or ribbon. Dried plants (petals and leaves), any mixture of abundant garden, wild, or weedy fragrant flowers. For wild plants, try roses, bee balm, purple coneflower, violets, fireweed, and geraniums.

HOW TO DO IT: Take the plant parts and mix it with the essential oil. Place it in a small cloth bag and tie with a piece of ribbon. Or use a square of cheesecloth and tie with ribbon (just make sure that all the corners are caught within the string). This makes a lovely, easy gift for a child to make and give away.

Note: For flower confetti to toss at a wedding, party, or celebration: Make the potpourri, break the dried materials into small pieces, and store in a cool, dry place until ready to use.

Layered petals

MATERIALS: Different-colored dried plants (keep the colors separate). Tall drinking glass or glass container with straight sides

HOW TO DO IT: Be sure that your dried plant material is separated by color. Take the material with the most abundant color and place it in the bottom of the glass to a depth of about 2 inches. Use a long wooden spoon to smooth out the top of the layer. Next, carefully put in a different color layer of material. Smooth out. Continue to layer the different colors.

Games

For a tic-tac-toe game, use two different kinds of flowers instead of X and O. For a game of checkers, instead of red and black pieces, use two different kinds of flowers (if you can find enough).

Scavenger hunt

Divide into teams. Each team has to find as many items on a list as possible. The team that finds the most—or finds them the fastest—wins. The list can be altered for different areas, different seasons, and different lessons. If you're teaching about different flower forms (umbel, simple, inflorescence etc.), you can add this to the list as well. This is just a basic list to get you started. I'd advise going to the site before finalizing your list so as to make sure everything is actually there.

As an alternative to picking things, ask the kids to take pictures instead.

1. Pink flower
2. Yellow flower
3. Blue flower
4. Vine
5. White flower
6. More than one flower on a stem
7. Feather
8. Some kind of cone from a tree
9. Seeds in a seedpod
10. Evidence of an animal (tracks, scat, half-eaten berries, and so forth)
11. Bird's nest
12. Grass
13. Evergreen leaf
14. Nut
15. Berry

APPENDIX

HELPFUL ORGANIZATIONS
NATIONAL

Lady Bird Johnson Wildflower Center (www.wildflower.org): Website includes Native Plants of North America, and offers information about more than 9,000 native plants.

US Forest Service (www.fs.fed.us/wildflowers): Good information about native plants. Includes links to all the state native plant societies and other related organizations.

Pollinator Partnership (www.pollinator.org): Promotes healthy plants and pollinators.

Audubon Society (www.audubon.org/native-plants): Provides a list of plants good for attracting birds specific to your zip code.

National Wildlife Federation (www.nwf.org/nativePlantFinder/plants): Provides lists of native plants (for your zip code area) that are good for attracting butterflies and moths.

STATE AND REGION

Almost every state has a Native Plant Society. For a list, go to the US Forest Service site (listed above). Several states and regions have outstanding organizations offering excellent websites:

www.illinoiswildflowers.info

www.missouribotanicalgarden.org

southwestdesertflora.com

montana.plant-life.org

www.nativeplanttrust.org (formerly called the New England Wild Flower Society)

www.wildflowersofcolorado.com

NATIVE PLANT NURSERIES

It's always best to acquire plants from nurseries as close to your garden as possible. An outstanding website, PlantNative, offers fairly complete lists of nurseries found in every state: www.plantnative.org/index.htm.

Lady Bird Johnson Wildflower Center (listed above) also offers suggestions for where to buy native plants.

CERTIFICATION PROGRAMS

Master naturalist: Most states offer a certification program for you to become a master naturalist. The EcosystemGardening website provides links to each state organization: www.ecosystemgardening.com/master-naturalist-programs-by-state.html.

Master gardener: Offered by state extension service personnel, this program provides training and information for you to become a master gardener. Most master gardener programs also provide good information about native plants. To find a program near you, check out the Extension Master Gardener website: mastergardener.extension.org/contact-us/find-a-program.

Backyard Wildlife Habitat certification: Offered through the National Wildlife Federation: www.nwf.org/garden-for-wildlife/certify.

INDEX

Page numbers in italics refer to illustrations.